GYMNOSPERMS OF INDIA AND ADJACENT COUNTRIES

K.C. SAHNI

Director (Retd.) Biological Research
Forest Research Institute, Dehra Dun

Formerly FAO/UNITED NATIONS Forest Botanist.

Illustrations by P.N. Sharma

BISHEN SINGH MAHENDRA PAL SINGH
23-A, Connaught Place
Dehra Dun-248 001
1990

GYMNOSPERMS OF INDIA
AND ADJACENT COUNTRIES

ISBN 81-211-0047-X

Printed by Gajendra Singh Gahlot at Shiva Offset Press,
and composed by Doon Phototype Printers, 14, Old Connaught Place, Dehra Dun
for Bishen Singh Mahendra Pal Singh, Dehra Dun, India.

Deodar forest Tehri Garhwal © Vincent James.

PREFACE

Gymnospermae is an extremely important Division of the Plant Kingdom. Economically they are of outstanding value, furnishing the greater proportion of our timber as well as resin. Most of the timber used in modern building is derived from conifers because of their straight grain and ease of manipulation. They are also the dominant forest makers of the world and in addition play a leading role in the preservation of the environment. Some of its members are resistant to pollution.

The region covered is vast, embracing eight countries of South Asia and S.E. Asia: India, Pakistan, Bangladesh, Nepal, Bhutan, Burma, Sri Lanka and the Maldives. This work up dates Gymnospermae by Sir Joseph Hooker in the fifth volume of the Flora of British India (1888). The present work covers the entire division Gymnospermae including the Gnetales. It also up dates Prof. M.B. Raizada and Prof. K.C. Sahni's *'Living Indian Gymnosperms'* Pt. I (1960) with a new format, with the latest classification. Gnetales was not covered in this work. Taxonomic and nomenclatural studies more particularly during the last two decades have necessitated changes in classification and in the use of many names. During this period there have been several expeditions to Arunachal Pradesh, Bhutan, Sikkim and Nepal which have resulted in the discovery of taxa new to science and new range extensions for several taxa. The author lead several expeditions to Arunachal Pradesh, Sikkim and the W. Himalaya collecting gymnosperms and recording field data which form the basis of this work. In the present work the family Taxaceae has been removed from the Coniferales and included in the new order Taxales. This was proposed by Professor Birbal Sahni and is being followed by modern Taxonomists. The Cephalotaxaceae has been retained in the Coniferales following the work of Professor Rudolf Florin. Some botanists place *Cephalotaxus* in the Taxaceae owing to its superficial resemblance with this family. However, the taxonomic position of *Cephalotaxus* is debatable.

New taxa are *Pinus bhutanica* first discovered by the author from Arunachal Pradesh in 1977 and later collected by British botanists from Bhutan in 1979; *Pinus wallichiana* var. *parva*, a new variety, with small female cones, seeds and shorter needles described by the author, also discovered from Arunachal Pradesh, *Larix himalaica*, a new larch described by Chinese botanists, endemic to S. Tibet and Nepal, *Picea farreri* Page and Rushforth a new find from Upper Burma. In addition a rare Taxaceae, *Amentotaxus assamica* Ferguson (1985) based on Kingdon Ward's collection of 1928 has been rediscovered after about 60 years from Lohit District, Arunachal Pradesh by K. Haridasan.

The present study has been undertaken to meet a demand from botanists and foresters of S. Asia. Seventeen genera and sixty species that are indigenous have been described in detail. In addition, common exotics have been briefly described. In the introduction, a brief account on phylogeny, changes in classification, distribution, etc. have been given. In subsequent pages, World Living Genera have been listed in an outline of classification with the indigenous appearing in bold print for quick reference, history of Gymnosperm classification, followed by sequence/arrangement of families, genera and synopsis of families.

The information on species includes, the latest valid botanical name with synonyms, with the date and place of publication of the name; a systematic and comprehensive botanical description; phenological data, to aid seed collection; details of distribution, including the citing of specimens. Information on associated animals and birds are recorded where such information was available. A note on propagation is provided, for each plant there is an illustration.

Acknowledgement is made to the undermentioned : to Shri P.N. Sharma for undertaking the exacting task of botanical illustrations based on herbarium material at FRI Herbarium, Dehra Dun; Mr. G.L. Lucas, Keeper, Herbarium Kew, for a photograph of *Juniperus turkestanica;* Dr. K. Thothathri, Joint Director, Botanical Survey of India, for loan of herbarium material of *Gnetum;* Sri B.M. Wadhwa, Indian Liaison Officer at Kew for information on *Ephedra* including photographs of this taxon; Dr. D.H. Nicolson, Smithsonian Institution, Washington, for supplying Dr. F.R. Fosberg's paper on Botany of the Maldives.

Dehra Dun
4 February, 1989

K.C. SAHNI

CONTENTS

List of photographs and silohouettes

List of illustrations

INTRODUCTION

This work covers the countries of the Indian sub-continent: Inaia, Pakistan, Nepal, Bhutan and Bangladesh. Burma, Sri Lanka and the Maldives are also covered.

The division Gymnospermae is one of great antiquity, reaching for back in geological history, at least two or three hundred million years. As a group they are interesting because of their antiquity, their excellent fossil record which is a nearly unbroken record of the past vegetation with a fascinating variety of genera and species which are in existence today. According to Tudge *et al.* (p. 173, 1988), 'during the carboniferous era, swampy low lands were dominated by the extinct, tree-like giant horse tails and early conifers. The conifers and their relatives (gymnosperms) subsequently rose in prominence but, during the cretaceous system (*c* 135 million years), they were increasingly replaced by broad leaved trees......' The total world living genera number sixty three and about 750 species. Seventeen genera and 60 species occur in our area including the recently discovered *Amentotaxus* from Arunachal Pradesh.

Although several of their members have become extinct, some like the Giant Redwoods of California are probably making their last stand along the Pacific coast of N. America and are vulnerable to extinction. As they are of global interest they are receiving special protection from foresters and conservationists. Gymnosperms particularly the coniferales are even today the dominant forest makers of the world, widely distributed in the temperate areas of both the northern and southern hemisphere and are an outstanding feature of the landscape of the Himalaya. They occur in all the continents except Antartica. The extinct gymnosperms seen as fossils today were contemporaneous with the extinct gigantic reptiles the dinosaurs which were ferocious looking giant sized lizards of the past.

The fossil Gymnosperms have been dealt with by the late Professor Birbal Sahni (1928, 1931) in his "Revision of Indian Fossil Plants" and later by A.R. Rao (1953) and others from the Institute of Palaeobotany, Lucknow.

Gymnosperms are one of the two great divisions of *Spermatophyta* or seed plants, distinguished from the Angiosperms by the fact that the ovules are always naked, not enclosed in an ovary, although they usually have some form of protection. Also, the endosperm (female prothallus) is formed before fertilization. There are two kinds of cones, male and female. The female cones are woody, with seeds borne on the upper sides of the scales. They are also characterized by mostly needle like leaves, exceptions being the Podocarps, *Agathis, Ginkgo,* etc. where they are broad-leaved. They are scale-like or awl shaped in *Cupressus, Juniperus, Calocedrus,* etc. The living Gymnosperms are divided into five orders *Cycadales, Ginkgoales, Taxales, Coniferales* and *Gnetales.* These orders differ very much from each other, are markedly distinct and are so widely divergent that they are now considered as example of parallel development. This view displaces the older view of lineal development, where by one family was interpreted to have been derived from another more primitive still existing family.

In recent years there have been changes in the classification of Gymnosperm orders. The most important being the removal of the family *Taxaceae* from the *Coniferales* and raising it to the status of an order *Taxales.* This was proposed by B. Sahni (1920b), and upheld by Florin (1948). Takhtajan (1953) also supported B. Sahni's classification and in addition retained his nomenclature. Harrison (1966, p. 1) in his revision of Dallimore & Jackson's *Handb. Conif. and Ginkgo.,* has said "There is an increasing body of opinion in favour of placing the family *Taxaceae* in a separate order the *Taxales".* The *Taxaceae* were elevated to the rank of an order by Pilger in 1916 a view he rejected 10 years later in 1926. (Pilger R. Taxaceae in Engler and Prantl, Die naturlichen Pflanzenfamilien ed 2, Bd 13: 199-211, 1926). The fact that the female reproductive organ in *Taxaceae* is not cone like but is characterized by a fleshy red aril enclosing a bony seed must have influenced Sahni to remove it from the order of "Cone bearers" the *Coniferales* and raise it to the order *Taxales* Sahni. This opinion is also accepted by Airy Shaw (1973, p. 1132) in his revision of Willis, *Dict. of Fl. Plants and Ferns.* It is interesting this reclassification proposed way back in 1920 came to be accepted by taxonomists much later and therefore has stood the test of time. For this reason the position of the other non 'cone bearers', the *Cephalotaxaceae* and *Podocarpaceae* may not be stable within the Coniferales where they have been placed since long Several other Gymnosperm orders are known only from fossils which occu as far back in geological history as the lower carboniferous period. Abou the time of the lower carboniferous period, two very distinct orders, th *Pteridospermales* or seed-bearing ferns, and the *Cordaitales* were evidentl already much diversified. By the end of the Permian or Triassic period however, both had declined and the surviving orders excluding the *Gnetales* of which there are few fossils had begun. The *Cycadales* are more closel related to the *Pteridospermales,* judging from their wood structure, leave and seed, whilst the *Coniferales, Taxales* and *Ginkgoales* resemble mor closely the *Cordaitales.* According to Airy Shaw (1973), 'The suggestio popular at one time, that some of the fossil Gymnosperms (*Bennettitales* with flower-like hermaphrodite cones might be on the direct line of ascent t the Angiosperms, is no longer given credence' (*see* Willis' *Dict. Fl. Pls. & Ferns,* rev. Airy Shaw ed. 8. p. 520, 1973).

Classification : The modern scheme of classification for Gymnosperm proposed by Melchior and Werdermann (1954) in Englers Syllabus de Pflanzenfamilien is now in vogue and is as under.

A. CYCADOPSIDA

1. *Pteridospermales :* Carboniferous, extinct.
2. *Bennettitales :* Mesozoic, extinct.
3. *Pentoxylales :* Jurassic.
4. *Cycadales :* Mesozoic to present day.

B. CONIFEROPSIDA

1. *Cordaitales :* Palaeozoic, extinct.
2. *Coniferales :* Palaeozoic to Mesozoic to present day.
3. *Taxales :* Jurassic to present day.
4. *Ginkgoales :* Palaeozoic to present day.

C. GNETOPSIDA

1. *Gnetales :* Recent.

The characteristics of these orders are as follows.

1. *Pteriodospermae* (*Cycadofolicales*). Trees or smaller plants. **Foliag**

fern-like, sporangia borne on frond-like leaves, more or less modified, not grouped into a cone.

2. *Cycadeoidales* (*Bennettitales*). Trees with stems covered with an armour of persistent leaf bases. Male sporophylls frond-like forming a loose crown. Female sporophylls not leaf-like, frond in specialized cone-like structure.

3. *Cycadales.* Plants with palm-like habit. Male and female sporophylls grouped into cones, except in *Cycas* where the female sporophylls are collected into a loose crown as in the *Cycadeoidales.* Male cells motile.

4. *Pentoxylales.* Low shrubs. Female infrutescence consisting of numerous cones spirally attached on a peduncle each bearing numerous ovules on a central axis with no megasporophylls, ovuliferous scales or interseminal scales. Male flowers terminally borne on dwarf shoots. Unilocular sporangia terminate short branchlets of the appendages or microsporophylls.

5. *Cordaitales.* Large trees with flat-strap-shaped leaves. Fructification in cones.

6. *Coniferales.* Large trees or shrubs. Leaves needle or scale-like, sometimes flattened. Male cells never motile. Male flowers arranged in more or less compact clusters or catkins. Female flowers usually clustered to form a cone.

7. *Ginkgoales.* Medium sized trees. Leaves flattened and lobed with dichotomous venation. Male and female elements collected into strobili.

8. *Gnetales.* Small trees (rarely) or shrubs or climbing shrubs. Opposite leaves, dicotyledonous embryos and compound cones in both male and female.

The existing Gymnosperms or the living Gymnosperms are divided into five orders : *Cycadales, Ginkgoales, Taxales, Coniferales* and the *Gnetales.*

Conifers the largest and most important order of the Gymnosperms do not bear flowers but produce pollen and seeds in cones and hence the name 'Conifer'. They are the simplest seed-bearing plants. Seeds and pollen are usually winged. A second distinguishing feature is that their leaves are narrow often needle-like (except *Podocarpus* and *Agathis*) where they are broad or lance-shaped. This reduces the surface area through which transpiration takes place and enables them to thrive under cold dry conditions. Economically they are the most valuable order of the Gymnosperms.

Almost ail are evergreen trees and some shrubs, shedding their old leaves and growing new ones not at one season but all the year round so that the tree is never bare of leaves. Indian Gymnosperms, are all evergreen, except the two Larches, *Larix griffithiana* confined to the E. Himalaya from E. Nepal to Arunachal Pradesh and *L. himalaica* endemic to Nepal and Tibet, which shed their leaves completely in winter, growing new leaves in the following spring.

Conifers are usually large and include some of the biggest and longest-lived of all trees. In the museum of the Forest Research Institute Dehra Dun, there is a 2.5 m wide section of Deodar more than 750 years old, dating back to the 13th century when Qutb Minar was built. This was growing in Tehri Garhwal near Balcha in Tons Division. There are some immense stands of deodar in Manali, Himachal Pradesh and in Garhwal near Mundali and Balcha. Redwoods of California, for example, are so huge that a tunnel wide enough for a motor-car to drive through has been cut through the trunk of one of them. The diameter of the trunk at ground level may reach 10 m. Considered to be the tallest tree in the world, trees up to 120 m/400′ high occur in the fog belt of the Pacific coast of N. America. Their age is estimated to be 3000 years. The Qutb Minar is 73 m high and the tallest deodar of 76 m would overtower it. It would be worthwhile introducing the Redwoods in the mist enshrouded Eastern Himalaya at 2000 m and at different altitudes.

Some trees of Foxtail Pine or Bristle-cone Pine (*Pinus aristata*) of N. America are estimated to be the oldest living things on earth today, having germinated even before the Indus civilization, some 6000 years (Sahni, K.C. 1986). The Sun Tree (*Chamaecyparis taiwanensis* Matsum) from Taiwan is claimed by scientists there to be more than 6000 years and measures 25 m/88′ girth (Prof. Chow Hui-Yen, International Wood Collectors Society 27(6) June 1974). It is growing 30 miles S.W. of Taipei and is claimed by Prof. Hui-Yen to be the oldest living tree in the world. Giant Sequoia (*Sequoiadendron giganteum*), a near ally of *Sequoia sempervirens* both classed as Redwoods is estimated to weigh 2000 tonnes and is the most massive tree in the world. The Great Banyan of Calcutta is a rival in its sheer bulk of tissue. Dawn Redwood (*Metasequoia glyptostro-*

boides Hù et Cheng) discovered in 1945 in Central China, was considered to have become extinct about 20,000,000 years ago. It was first discovered as a fossil in Japan in 1941 by Miki. The discovery of living trees aroused world wide interest, and to save it from extinction seeds were distributed to many Botanical Gardens and trees are now flourishing in Darjeeling and Almora. It was raised in the Botanical Garden of the Forest Research Institute, Dehra Dun where it died after 5 years. As its popular name 'Water Fir' or 'Water Larch', suggests it ought to be tried this time along the water course within the Botanical Garden of the Forest Research Institute. In its natural habitat in E. Szechwan China at 700-1350 m it grows along streams. The climate of Dehra Dun and the eastern Himalaya are suitable for its growth and an altitude of 2000 m in the Eastern Himalaya should be suitable. It has potential in forestry because of its fast rate of growth upto 1 m a year. At the same time it is ornamental with graceful foliage turning coppery red in the autumn. Dawn Redwood, as well as *Ginkgo biloba* introduced in the Himalaya are designated as living fossils.

The conifers are entirely wind fertilised, the pollen is light and powdery. Sometimes provided with air-bladders (*Pinus*), it is produced in enormous quantities called 'pollen showers' resembling sulphur dust. About the time it is shed, the scales of the female cones open to receive it and the pollen grains alight on the scales and adhere to the sticky fluid at the apex of the ovule. Fertilization often does not take place for a long time afterwards. The seeds in many genera with woody cones (*Pinus, Abies,* etc.) are winged for wind dispersal; in other genera they are animal distributed, e.g. *Juniperus* where the cone is fleshy.

Distribution

The genera of Northern Hemisphere Gymnosperms which occur in our area are the *Coniferales* (*Pinus, Abies, Picea, Tsuga, Cedrus, Larix, Cephalotaxus, Podocarpus, Taiwania, Juniperus, Cupressus, Calocedrus,* popularly called Northern Incense Cedar), *Cycadales (Cycas), Taxales (Taxus, Amentotaxus), Ginkgoales* (*Ginkgo* introduced), *Gnetales (Gnetum* and *Ephedra).*

The Southern Hemisphere Gymnosperms mostly introduced in our area are : *Araucaria, Agathis,* and *Podocarpus.* The last named, *Podocarpus* occurs within the Southern and Northern Hemisphere. Podocarps occur naturally in our area in E. India, E. Himalaya, Andaman and Nicobar

Islands, S. India and Burma. The southern hemisphere is deficient in Conifers. However the introduction of Radiata Pine *Pinus radiata* of Monterey, California have been outstandingly successful in Australia and New Zealand along with Monterey Cypress (*Cupressus macrocarpa*). A detailed general account on distribution is dealt with in Raizada and Sahni (1960) and Biswas (1933). In the present account the most outstanding features on distribution are highlighted coupled with new findings of new species and new range extensions of taxa reported since the publication of Raizada and Sahni (1960). However, detailed distribution in tabular form is given for quick reference, including distribution of *Gnetales* updating Raizada & Sahni (loc. cit.), with new species, new range extensions. Detailed authentic distribution is given under species accounts.

The only native conifer of S. India is *Podocarpus wallichianus* C. Presl from the Nilgiris. This has disjunct distribution occurring in the far off Khasi Hills, Burma and collected for the first time from the Great Nicobar Island by the author (Sahni, 1953).

Sri Lanka is totally devoid of native conifers and *Gnetales*, both *Gnetum* and *Ephedra*. The only gymnosperm in the Maldives is *Cycas circinnalis*. Although the conifer forests of the Western Himalaya are far more extensive and gregarious, in the Eastern Himalaya they are not that vast. Here moisture loving *Lauraceae* and oaks occupy large areas. However Gymnosperms exceed in the number of genera and species in the E. Himalaya with new elements coming in like, *Larix griffithiana*, *L. himalaica*, *Tsuga dumosa* (from E. Kumaon eastwards), *T. yunnanensis*, *Amentotaxus assamica*, *Podocarpus*, *Cycas*, *Gnetum*, *Taiwania cryptomerioides*, *Calocedrus macrolepis*, etc., the last two and *Tsuga yunnanensis* occurring in the Burmese mountains which are a continuation of the Himalaya. The Eastern Himalaya is so rich in Gymnosperms that just one state of Arunachal Pradesh has five species of *Pinus* as compared to Uttar Pradesh which have only two, viz. 'Chir' pine and 'blue pine'. The Arunachal Pines are, *P. roxburghii*, *P. wallichiana*, *P. bhutanica* (new find), *P. merkusii*, and *P. armandii*. Chilgoza pine *P. gerardiana* is endemic in the N.W. Himalaya from Kinnaur westward to Kashmir, Chitral, N. Baluchistan, and Afghanistan at 1800-3000 m in the inner semi-arid valleys where precipitation is in the form of snow with scanty rainfall. *Cedrus deodara* is also endemic to N.W. Himalaya, Pakistan Himalaya, Kashmir,

Himachal, Garhwal, absent in Kumaon, present in Kurnauli valley in W. Nepal.

Pinus bhutanica a new species was collected by the author from Tenga Valley, Kameng District, Arunachal Pradesh in April 1977 and later in 1979 by Grierson and Long from Bhutan. Haridasan (1988), botanist of the Arunachal Forest Department rediscovered recently a most rare *Taxales, Amentotaxus assamica* from Delie Valley, Lohit District in Arunachal Pradesh identified at Herbarium FRI, Dehra Dun. It was first collected by Lister from Dafla Hills in 1875 and later by Kingdon Ward in 1928 from Delie Valley. Its rediscovery after 60 years is interesting. Since it is endemic and extremely rare, conservation measures are necessary to prevent it from extinction. The two Needle Pine, *Pinus merkusii* of Burma has been collected in a wild state in the extreme eastern limit of Lohit District where it occurs in quantity and is regenerating very well. It is the most tropical of all pines and is the only pine in the world that crosses the equator. Besides Burma it occurs in Thailand, Sumatra, Cochin-China, Borneo and the Philippines. The range of distribution of *Podocarpus neriifolius* has been extended to Kameng District, Arunachal Pradesh by its recent collection by Sahni & Naithani (No. 559, Herbarium FRI, Dehra Dun). It also occurs in Sikkim, Bhutan and the Andamans. A spruce *Picea brachytyla* was collected from Kameng District (Sahni, 1962). This added a conifer new to India. Recently found in Bhutan in Bhumthang (Grierson & Long, 1983). It also occurs in W. Hupeh and W. Szechwan in China. Page and Rushforth (1980) described a new taxon *P. farreri* collected by Farrer from 'The Valley of Winds and Water' in Upper Burma.

The W. Himalaya has 2 species of fir, the low level *Abies pindrow* and the high level *A. spectabilis.* The E. Himalaya has 3, *A. spectabilis,* the umbrella or pagoda-shaped *A. densa* which is widely distributed and *A. delavayi* that is scarce confined to small areas in Kameng District. According to Thierot (1980, p 520), 'Gymnosperms grow at almost the highest altitude reached by vascular plants. A fir (*A. spectabilis*)* has been found at 5,350 m/c 17,500′ in Kashmir'. The author so far has not seen any tree at this altitude in Kashmir or the W. Himalaya.

---

* The name in parenthesis by the author.

The most interesting Coffin Wood Tree *Taiwania cryptomerioides* occurs in N. Burma, S.E. Tibet, S.W. Yunnan and Taiwan. It is carried overland to China from N. Burma where it is in demand as coffin boards by wealthy Chinese because it is durable and strong. The discovery of the Northern Incense Cedar *Calocedrus macrolepis* from Southern Shan States at 2150-2400 m is an interesting discovery in Burmese Flora.

Polunin & Stainton (1984) report *Larix himalaica* W.C. Cheng & L.K. Fu from Nepal and consider it to be endemic to C. Nepal and Tibet. This new find described by Chinese botanists adds another Larch to the Eastern Himalaya. (see *Larix*).

A stimulating account of the geography of conifers and taxads, past and present, is given by Prof. Rudolf Florin (1963) in *Acta Horti Bergiani* **20**(4): 121-312.

According to Theirrot (loc. cit.) 'Gymnospermous trees reach more northerly latitudes than do angiospermous trees. On the fringes of the Arctic (i.e., along the continental "tree line") in North America and Eurasia are only two kinds of trees, spruces and larches, both gymnospermous'.

Distribution of Gymnosperms in India and adjacent countries.

Name	No. of species in the region covered	Family	Distribution
1	2	3	4
Cycas.	5+1 var.	*Cycadaceae*	India, Bangladesh, Bhutan, Burma, Nepal, Sri Lanka and the Maldives.
C. beddomei		"	Cuddapah District (Andhra Pradesh), Tamil Nadu, Endemic.
C. circinnalis		"	W. Ghats, Malabar, Orissa, Sri Lanka and Maldives; sea level to 1070 m′.
C. circinnalis var. *orixensis*		"	Puri and Angul (Orissa).

1	2	3	4
C. pectinata		"	Sikkim, Nepal, Sameshwar Hills, Bihar (rare), Assam (Goalpara, Kamrup) Khasi Hills, Manipur Chittagong, Martaban and Tenasserim in Pine forests.
C. rumphii		"	S. Burma, Cocos Islands, Andamans, Nicobars, Sri Lanka, Malaya, New Guinea, and N. Australia, littoral.
C. siamensis		"	Burma, Thailand, Cochin-China and Yunnan.
Taxus	1	*Taxaceae*	All along the Himalaya from Pakistan to Arunachal Pradesh, through Nepal, Sikkim, Bhutan, Khasi Hills, Manipur and Burma.
T. baccata		"	All along the Himalaya from Pakistan-Arunachal Pradesh. Khasi Hills, Naga Hills. Manipur and Burma. 1800-3000 m.
Amentotaxus	1	"	Arunachal Pradesh, other spp in C & S China, Taiwan, Vietnam.
A. assamica		"	Delei Valley, Lohit District, Arunachal Pradesh. Endemic.
Cephalotaxus	2	*Cephalotaxaceae*	Khasi, Jaintia and Naga Hills, Arunachal Pradesh.
C. mannii		"	Khasi, Jaintia and Naga Hills from 1370-2590 m.
C. griffithii		"	Kameng District, Arunachal Pradesh, Naga Hills, Manipur, Mishmi Hills at 1520-1820 m and in Burma.
Podocarpus	4	*Podocarpaceae*	Burma, Andamans, Gt Nicobar, Sikkim, Bhutan, Arunachal Pradesh, Nilgiris.
P. imbricata		"	Upper Burma (800-1000 m), China, N Borneo, Java, New Guinea, Philippines.

1	2	3	4
P. neriifolius		"	Sikkim, Bhutan, Kameng District (Arunachal Pradesh), Andamans, Burma, SW China, Malaya Peninsula. Sunda Islands.
P. wallichianus		"	Western Ghats from Nilgiris southwards, Great Nicobar, Assam, Martaban and Tenasserim.
P. macrophyllus		"	N Burma.
Pinus	7 + 1 var.	*Pinaceae*	Afghanistan, all along the Himalaya, Meghalaya, Naga Hills, Manipur, Burma, China, Taiwan, Thailand, Sumatra, Java, Borneo and the Philippines.
P. wallichiana		"	All along the Himalaya from Pakistan-Arunachal Pradesh 1800-3700 m. Rare in Sikkim and a considerable portion of Kumaon.
var. *parva*			Tawang, 10,000′ Arunachal.
P. armandii :	Lohit Dt. 7000′ (Kingdon Ward)	"	Arunachal Pradesh, Tseng Po Gorge, Trulung SE Tibet, NE Frontier (Burma) C & W China, Taiwan.
P. bhutanica		"	Tenga Valley in Kameng District, Arunachal Pradesh and Bhutan.
P. gerardiana		"	Bashahr westwards to Kashmir, Chitral N Baluchistan and Afghanistan 1800-3000 m.
P. kesiya		"	Khasi and Naga Hills, Manipur, Upper Burma and the Philippines 800-2000 m.
P. roxburghii		"	All along the Himalaya from Pakistan to Arunachal Pradesh at 450-2300 m, except the Kashmir Valley Proper.

1	2	3	4
P. merkusii		"	Lohit District (Arunachal Pradesh), Burma, Thailand, Cochin-China, Sumatra, Java, Borneo and the Philippines.
Tsuga	2	"	E Kumaon eastward to Arunachal Pradesh, Upper Burma, China.
Tsuga dumosa		"	E Kumaon to Arunachal Pradesh, Upper Burma.
T. yunnanensis		"	W Yunnan to W Szechuen 2130-3048 NE upper Burma.
Picea	4	"	All along the Himalaya, N Burma (rare) W Szechuen (China).
P. smithiana		"	In the W Himalaya, from Afghanistan to Kumaon at 2150-3300 m.
P. spinulosa		"	Sikkim, Bhutan to Arunachal Pradesh 2600-3350 m.
P. brachytyla		"	Kameng District (Arunachal Pradesh), Bhutan in Bumthang, W Hupeh & W Szechwen, 1500-2300 m.
P. farreri		"	Upper Burma. (Fen-Shui-Ling Valley).
Abies	4	*Pinaceae*	All along the Himalaya up to Arunachal Pradesh. NE Burma, China.
A. pindrow		"	Afghanistan, Pakistan Himalaya-Nepal (W Himalaya-Nepal) at 2300-3350.
A. spectabilis		"	All along the Himalaya from Pakistan-Arunachal Pradesh, Tibet 2800-4300 m. Commonly between 3300- 4000 m.
A. densa		"	E Nepal, Darjeeling to Arunachal Pradesh, NE Frontier (Burma) at 2750-3950 m.

1	2	3	4
A. delavayi		”	Peri La also above Tawang, Se La (Kameng District, Arunachal Pradesh), Bhutan, NE Upper Burma, SW Szechuen (China) 3000-3300 m.
Cedrus	1	”	Afghanistan to Garhwal, Kurnauli Valley (W Nepal), 1200-3300 m.
C. deodara		”	Afghanistan, Kashmir, Himachal, Garhwal, Kurnauli Valley (W Nepal) 1200-3300 m.
Larix	2	”	E Nepal-Arunachal Pradesh, Chumbi Valley (Tibet), NE Upper Burma, 2400-3700 m. Common from 2900-3300 m.
L. griffithiana		”	E Nepal-Arunachal Pradesh, Chumbi Valley (Tibet), N E Upper Burma, 2400-3700 m. Common from 2900-3300 m.
L. himalaica		”	C Nepal and Tibet 2400-4000 m. Endemic.
Taiwania	1	*Taxodiaceae*	N Burma, Yunnan and Taiwan.
T. cryptomerioides		”	N Burma, Yunnan and Taiwan at 1800-2600 m.
Juniperus	7	*Cupressaceae*	N Hemisphere from the Arctic circle to Mexico and W Indies, Azores, Canary Isls, N Africa, E trop. Afr. (mountains), Himalaya, China and Japan.
J. polycarpos		”	Afghanistan, Baluchistan, Kagan Valley, Kashmir, Lahul to Kumaon, W Tibet, 2500-4300 m.
J. recurva		”	All along the Himalaya in inner valleys to Arunachal Pradesh, Upper Burma, China. 2700-4600 m.
J. recurva var. *coxii*		”	E Himalaya (Tsangpo gorge Kingdon Ward), N Burma.

1	2	3	4
J. communis		"	Kumaon westwards to Afghanistan, temperate and subarctic Europe Asia, N Africa and N America.
J. squatmata		"	Afghanistan, Himalaya, NE Burma, China, Taiwan and Japan.
J. indica		"	All along the Himalaya, from Indus to Sikkim, Bhutan and W Tibet.
J. turkestanica		"	Nanga Parbat, Kazakhistan and Uzbekistan (C Asia), Pamirs.
Cupressus	3	*Cupressaceae*	Gurez (Kashmir), Himachal Pradesh to Nepal, Arunachal Pradesh (uncommon) Ladakh? China, Uttar Pradesh, China.
C. torulosa		"	Chamba (Himachal) to Nepal, W Szechuen.
C. corneyana		"	Bhutan, Arunachal, Tibet.
C. cashmeriana		"	Unknown in wild, Ladakh?
Calocedrus	1	"	Burma, Taiwan, Yunnan.
C. macrolepis		"	S Shan States (Burma) S W Yunnan and Taiwan. 2150-2400 m.
Gnetum	Gnetaceae absent from Sri Lanka. 8 species, 4 varieties and two formas	"	W Ghats, Eastern Coast and Ghats including Orissa, W Bengal, Sikkim, Bangladesh, Arunachal Pradesh, Assam, Nagaland, Manipur, Mizoram, Andaman and Nicobar Islands and Burma. None in Sri Lanka.
G. gnemon			The tree form in our area is only found in the Great Nicobar Island where it was reported by Thothathri *et al.* (1976). Elsewhere it is found in Indonesia which is adjacent to the Great Nicobar Island.

1	2	3	4
G. gnemon var. brunonianum			Assam (Golaghat) S Mizoram
G. gnemon var. *griffithii*			Assam (Sibsagar), Naga Hills, Kungba 1500 m. Burma (Tavoy, Tenassrim River, S. Tenasserim, Nagawun, Chaung forest, Mergui) Ascends up to 1500 m in the rainforests in Naga Hills and Mizoram.
G. oblongum			Bangladesh (Chittagong, Demagiri), Burma (Tonkyghat, Tavoy, Wazum Chaung, Mergui, Lemji).
G. ula			Maharashtra (Yellapur, Khandala Valley, Poona). Goa, Kanara, Karnataka (Coorg), Kerala (Nilambar, Quilon), Tamil Nadu (Nilgiris, Devamalai) at 1300 m in Nilgiris, Pulni Hills, Chenath Nair Hills, Andhra Pradesh (Godaviri Distt), Orissa (Ganjam Mahendragiri), S Andamans.
G. montanum			Darjeeling, Sikkim at 1500 m, Bangladesh (Silhet) Assam, Orissa (Mayurbhanj), Burma (Katha Distt.) Pegu, Tharra waddy, Martahan.
G. montanum f. megalocarpum			Arunachal Pradesh (Dafla Hills). *Lister.*
G. contractum			Tamil Nadu (Conoor in Nilgiris), Kerala (Quilon) up to 1500 m.
G. latifolium var. *macropodum*			Andamans (Port Blair, Mt. Harriet), S. Andamans. Nicobars (Kamorta).
G. latifolium var. *funiculare*			S. Andamans (Port Monat).
G. macrostachyum			Burma (Tavoy)
G. microcarpum f. silvastre			Burma (Mergui)
Ephedra	8 spp. 3 vars.		

Tribe *Tropidolepides*

E. przewalskii Stapf : Nanga Parbat (Kashmir), along bed of Shyok-River in Ladakh.

Tribe *Scandentes* Stapf

E. foliata Boiss : Baluchistan, Sind, Rajasthan, Punjab.

Tribe *Pachycladae* Stapf

E. pachyclada Boiss : Chitral, Baluchistan, Kurram Valley.

E. intermedia Schrenk et C.A. Mey var. *tibetica* Stapf : Lahul, Spiti, Pangi, Rowli Chamba, Ladakh.

Tribe *Leptocladae*

E. regeliana Florin : Karokorum, Ladakh.

E. gerardiana Wall. ex Stapf : Karakorum, Chitral, Gilgit, Baltistan, Hazara, Kashmir, Ladakh, Zanskar, Kistwar, Chamba, Lahul, Garhwal, Kumaon, Badrinath.

E. saxatilis (Stapf) Royle ex Florin, Simla, Jaunsar (Garhwal) Badrinath Kunawar, Kumaon, S. Tibet.

var. *sikkimensis* (Stapf) Florin : N Sikkim (Chho Lhamo 17000′), S Tibet.

E. nebrodensis Tin. var. *procera* (Fisch et Mey) Stapf, Baluchistan, Kashmir, Chamba, Lahul.

Economic importance

Economically they are the most important, furnishing the greater proportion of our timber, as well as resins, tars, turpentines, etc. Most of the timber used in modern building is derived from conifers because of their straight grain and ease of manipulation. Notable as source of pulpwood for paper manufacture are pines, firs and spruces. Spruce and pine are sources of oils used as scents in soap, air freshners and perfumes. Cones of pine and other conifers are sources of food for wildlife providing nutritious seeds for birds in winter (Tudge 1988 p. 103, p. 176) Chilgoza seeds are popular as dry fruits during the winter. Conifer forest are typical of the Himalaya and provide a cool and soothing environment for recreation and health. They provide material for landscaping of parks and gardens. Extinct fossil gymnosperms contributed greatly to the formation of coal. Most of the workable coal seams of the Peninsular Lower Gondwanas of India contain a significant proportion of gymnospermous plant remains. (Thierot 1980; Sahni K.C. Conifers, In. *Encyclop. Ind. Nat. Hist.* p 134, 1986).

WORLD LIVING GENERA & FOSSIL ORDERS

An outline of classification. (Genera in bold print are native to S. Asia)

Division	GYMNOSPERMAE	
1. Class	CYCADOPSIDA	
1. Order	Pteridospermae	Carboniferous Extinct.
2. Order	Caytoniales	Only known from fossils.
3. Order	Cycadales	Mesozoic to present day : Dioon, Ceratozamia, Microcycas, Zamia (W Hemisphere). **Cycas**. Macrozamia, Encephalartos, Stangeria, Bowenia (E Hemisphere).
4. Order	Nilssoniales	Only known from fossils.
5. Order	Bennettitales	Only known from fossils.
6. Order	Pentoxylales	Only known from fossils.
2. Class	GINKGOPSIDA	
7. Order	Ginkgoales	
Family	Ginkgoaceae	Ginkgo & fossil members. Palaeazoic-present day.
3. Class	CONIFEROPSIDA	
8. Order	Cordaitales	Only known as fossils.
9. Order	Coniferales	Araucariaceae, Cephalotaxaceae, Cupressaceae, Pinaceae, Podocarpaceae, Taxodiaceae.
Family	Lebachiaceae, Voltziaceae, Cheirolepidaceae and Protopinaceae	Only known as fossils.
Family	Pinaceae	

Subfamily	Abietoideae	**Abies**, *Keteleeria*, *Cathaya*, *Pseudotsuga*, **Tsuga**, **Picea**
Subfamily	Laricoideae	*Psuedolarix*, **Larix**, **Cedrus**
Subfamily	Pinoideae	**Pinus**
Family	Taxodiaceae	(a) *Sequoia*, *Sequoiadendron* (b) *Metasequoia* Miki ex Hu et Cheng (c) *Taxodium*, *Glyptostrobus* (d) *Cryptomeria* (e) *Cunninghamia* (f) *Sciadopitys* (g) *Athrotaxis*, **Taiwania**
Family	Cuppressaceae	
Subfamily	Callitroideae	
Tribe	Actinostrobeae	*Actinostrobus*, *Callitris*, *Fitzroya*
Tribe	Libocedreae	*Neocallitropsis*, *Widdringtonia*, *Diselma*, *Papuacedrus*, *Austrocedras*, *Libocedrus*, **Calocedrus**
Tribe	Tetraclineae	*Tetraclinis*
Subfamily	Cupressoideae	
Tribe	Cupresseae	**Cupressus**, *Chamaecyparis*, *Cupressocyparis*, *Fokienia*
Tribe	Thujopsideae	*Thujopsis*, *Thuja*
Tribe	Junipereae	**Juniperus**, *Microbiota*
Family	Podocarpaceae	
Subfamily	Pherosphaeroideae	*Microstrobos*
Subfamily	Phyllocladoideae	*Phyllocladus*
Subfamily	Podocarpoideae	*Saxegothaca*, *Microcachrys*, *Dacrydium*. **Podocarpus**, *Acmopyle**
Family	Cephalotaxaceae	**Cephalotaxus**
Family	Araucariaceae	*Agathis*, *Araucaria*
4. Class	TAXOPSIDA	
10. Order Family	Taxales B. Sahni Taxaceae	**Amentotaxus**, Pseudotaxus Cheng syn Nothotaxus Florin,

*In view of Birbal Sahni's contributions to Gymnospermae, *Acmopyle sahniana* Buchh.-Gray is named after him from Fiji (Dallimore & Jackson, p. 91, 1966).

		Austrotaxus, Torreya, **Taxus** (1, 2, comparatively recently discovered)
5. Class	CHLAMYDOSPERMAE Pulle 1938 (CHLAMYDOSPERMOPSIDA)	
11. Order	Gnetales	
Family	Welwitschiaceae	*Welwitschia*
Family	Ephedraceae	**Ephedra**
Family	Gnetaceae	**Gnetum**

Summary of History of Classification :

Bentham & Hooker (1862-1883) placed the gymnosperms between the Dicotyledones and the Monocotyledones. Hofmeister's (1851) remarkable investigation on the embryology of plants prompted Van Tiegham (1898) to remove them from this intermediate position and instal them as one of 2 divisions of the Spermatophyta : the GYMNOSPERMS and the ANGIOSPERMS. The discovery of motile sperms in Ginkgo by Hirase (1895, 1896, 1898) led Engler (1897) to create a new order Ginkgoales. Engler & Prantl (1926) divide the Gymnosperms into seven classes or orders (1) Pteridospermae (Cycadofilicales) (2) Cycadales (3) Bennettitales (4) Cordaitales (5) Ginkgoales (6) Coniferales (7) Gnetales.

B. Sahni (1920a) and (1920b) recognized 2 phyletic lines in the Gymnospermae. (1) Phyllosperms consisting of (a) Pteridospermae (b) Cycadales (c) Bennettitales and (2) Stachyosperms having (a) Cordaitales (b) Ginkgoales (c) Coniferales (d) Taxales (which he separated from the Coniferales and included *Taxus, Torreya* and *Cephalotaxus*). (*Amentotaxus* had not been discovered at that time). As their names indicate, the ovules of Phyllosperms are borne on leaves, those of Stachyosperms are stem borne.

The outstanding feature of Sahni's classification that has come to stay is the separation of *Taxus* and its allied genera from the Coniferales and their inclusion in a new order, *Taxales* Sahni (*see* Willis, 8th edn by Airy Shaw 1973). Florin (1948) upheld Sahni's separation of the Taxales as an order equal in rank with the Coniferales, he included *Taxus, Torreya* and the three more recently discovered genera viz. *Pseudotaxus* Cheng syn *Nothotaxus* Florin, *Amentotaxus* and *Austrotaxus* and their fossil relatives in the order, retaining *Cephalotaxus* within the Coniferales. Pulle (1938) included the Gnetales in a separate class, the Chlamydospermae, equal in rank with his other classes, viz Pteridospermae, Gymnospermae and Angiospermae. Takhtajan (1953) retained Sahni's nomenclature and Lam (1950) extended the distinction even to Angiosperms.

Engler, Melchior & Werdermann (1954) proposed another treatment. (1) Gymnospermae is treated as a division of the plant kingdom equal in rank with Bryophyta, Pteridophyta, Angiospermae *et al.* (2) The nomenclature of Chlamydospermophyta is changed to Chlamydospermae. The 3 genera, *Ephedra, Welwitschia* and *Gnetum* are retained in a single order, Gnetales. (3) The Caytoniales have been separated from the Pteridospermae and Nilssoniales from the Cycadales and introduced as a distinct order in the class CYCADOPSIDA. (4) A new order Pentoxylales including the Pentoxyleae (Srivastava 1945; B. Sahni 1948) is incorporated in the CYCADOPSIDA. The Pentoxyleae is a unique group from the Rajmahal Hills which in Sahni's words "defies classification and presents a new problem in our understanding of the evolution of gymnosperms." The group is remarkable for its peculiar combination of characters suggestive of affinities with the Cycadales, the Cycadeoidales and the Coniferales. (5) The Ginkgoales have been removed from the CONIFEROPSIDA and included in the CYCADOPSIDA and recently raised to a new class GINKOPSIDA in *Fl. Reipublicae Popularis Sinicae, Gymnospermae*, Beijing (1978).

Pant (1957) put forward a new classification which is very briefly tabulated below. (*see* Pant *loc. cit.* for a detailed treatment).

Division 1. Cycadophyta

- Class 1 Pteridospermopsida
- Class 2 Cycadopsida
- Class 3 Pentoxylopsida
- Class 4 Bennettitopsida

Division 2. Chlamydospermophyta-peculiar angiosperm-like forms.

- Class 1 Gnetopsida

Division 3. Coniferophyta-usually large-sized trees.

- Class 1 Coniferopsida
- Class 2 Ephedropsida
- Class 3 Czekanowskiopsida
- Class 4 Taxopsida

GYMNOSPERMAE Lindl.

Trees, shrubs rarely climbers (*Gnetum* and *Ephedra*). Characterized for the most part by xeromorphous acicular (*Pinaceae*) or fern-like foliage (Cycads) and unisexual inflorescence. Ovules naked, not enclosed in an ovary, but lie exposed on the carpels which are open and therefore devoid of a style and stigma. Pollen grains spherical or oval, often with bladder-like extension of the exine, and containing a prothallus of two or more cells, one of which produces two non motile or motile male gametes or spermatozoides (*Ginkgo, Cycas*). Pollen grains usually wind borne settling directly upon the micropyle. Embryo sac before fertilisation contains a tissue which can be termed a rudimentary prothallium and in it are formed two or more archegonia, each consisting of a large egg cell and a short neck. Only one embryo usually develops to maturity. It consists of an axis, bearing two to several cotyledons and ending in a radicle, the whole being enclosed in a mass of endosperm which represents a continuation of the gametophytic tissue. The sporophylls are usually arranged in a cone-like structure.

Considering the differences between the arrangement of the reproductive organs in the Angiospermae and the Gymnospermae, the term 'strobilus' (plural strobili) is used in this work for Gymnosperm 'flowers'.

THE SEQUENCE AND ARRANGEMENT OF ORDERS, FAMILIES AND GENERA

Division	GYMNOSPERMAE
ORDER I	CYCADALES
FAMILY 1	CYCADACEAE (Cycad Family)
Genus (1)	*Cycas*
ORDER II	*GINKGOALES
FAMILY 2	GINKGOACEAE (Ginkgo Family)
Genus (1)	Ginkgo
ORDER III	TAXALES
FAMILY 3	TAXACEAE (Yew Family)
Genera (1)	*Taxus*
(2)	*Amentotaxus*

ORDER IV	**CONIFERALES**
FAMILY 4	**CEPHALOTAXACEAE** (Plum-Yew Family)
Genus (1)	*Cephalotaxus*
FAMILY 5	PODOCARPACEAE (Podocarp Family)
Genus (1)	*Podocarpus*
FAMILY 6	*ARAUCARIACEAE (Monkey-Puzzle Family)
Genera (1)	*Agathis
(2)	*Araucaria
FAMILY 7	PINACEAE (Pine Family)
Genera (1)	*Pinus*
(2)	*Tsuga*
(3)	*Picea*
(4)	*Abies*
(5)	*Cedrus*
(6)	*Larix*
FAMILY 8	TAXODIACEAE (The Red wood Family)
Genera (1)	*Taxodium
(2)	*Cryptomeria
(3)	*Taiwania*
FAMILY 9	CUPRESSACEAE (Cypress Family)
Genera (1)	*Juniperus*
(2)	*Cupressus*
(3)	*Callitris
(4)	*Thuja
(5)	*Calocedrus*
ORDER V	GNETALES

*Exotic to the region covering this work.

FAMILY 10	GNETACEAE (Gnetum Family)
Genus	*Gnetum*
FAMILY 11	EPHEDRACEAE (Ephedra Family)
Genus	*Ephedra*

SYNOPSIS OF FAMILIES

The distinguishing characters of the eleven families are selected with special reference to the trees and shrubs described in this work.

1. CYCADACEAE Trunk cylindric, sometimes branched, with terminal tufts of rigid, pinnate leaves. Male and female reproductive organs on separate plants, at the apex of trunk among the leaves, male in erect massive cones with thick scales, bearing on the undersurface indefinite globose anther cells. Female carpophylls (megasporophylls) densely wooly in crowded whorls around the top of the trunk, each with pairs of large ovules on its edge.

2. GINKGOACEAE Monotypic trees. Leaves deciduous, fan shaped, with parallel veins. Dioecious. Male strobilus catkin-like; antherozoids motile. Female axis a long stalk with usually 2 terminal ovules. Fruit drupe-like, orange yellow, woody shell of the seed surrounded by a fetid, pulpy coat.

3. TAXACEAE Evergreen trees. Leaves spirally arranged, rarely opposite, flattened, needle-like, linear, often in 2 ranks, male strobili catkin-like. Female not cone-like, with a single erect ovule or seed, partly or wholly covered by a fleshy, coloured aril. Foliage and seeds poisonous, but not the aril.

4. CEPHALOTAXACEAE Evergreen trees or shrubs. Branches opposite whorled. Leaves spirally arranged, often spreading in 2 ranks, linear, pointed. Usually dioecious. Male strobili axillary, sub-globose, sporophylls with 3-8 pollen-sacs. Ovules in pairs, only one ovule developing into an olive-like 'fruit' (seed), long stalked, with a fleshy outer coat.

5. PODOCARPACEAE Evergreen trees or shrubs. Leaves spirally arranged, rarely opposite, scale-like, needle-like linear or lanceolate or

oblanceolate. Monoecious or dioecious. Male strobili catkin-like, microsporophylls with 2 pollen-sacs. Ovules inverted; outer seed-coat dry or fleshy in reduced cones or supported by a fleshy bract and stalk.

6. ARAUCARIACEAE Trees with spirally arranged narrow or broad leaves with parallel veins. Dioecious or monoecious. Male strobili dense, cylindrical with numerous sporophylls. Cones sub-globose to ovoid, generally large; with woody, one-seeded scales, without distinct bracts, falling when the seeds are ripe.

7. PINACEAE Trees. Leaves linear, flat or needle-like, spirally arranged sometimes in 2 or more rows, or in tufts. Buds scaly. Monoecious. Cones mostly woody, with spirally arranged scales. Bracts separate from the scales, except at their base, generally small and relatively narrow, sometimes longer than the scales (disappearing in the mature cone in *Pinus*). Seeds 2 to each fertile scale inverted, generally winged.

8. TAXODIACEAE Trees with narrow, linear or awl-shaped leaves, spirally arranged, sometimes apparently 2 ranked. Buds not scaly. Monoecious Cones mostly globose. Scales spirally arranged, leathery or woody, persistent, more or less fused with bracts. Seeds 2-6 to each scale, erect or inverted, winged or wingless.

9. CUPRESSACEAE Trees or shrubs, adult leaves usually small, scale-like, appressed, in opposite pairs, sometimes in threes, rarely linear. Buds not scaly. Cones small their scales mostly confluent with the bracts, somewhat fleshy but generally woody when ripe, each bearing a number of erect ovules.

10. GNETACEAE Scrambling climbers, very rarely trees; leaves opposite elliptic or ovate, venation pinnate; 'flowers' whorled on spike like axes. Fruit drupe-like, seeds hard ellipsoid, with a fleshy integument.

11. EPHEDRACEAE Shrubs, with one climber (*E. foliata*), bearing closely packed green stems with obscure deciduous scale-like leaves; 'flowers' in small axillary fascicles; seeds ovoid 1-2.

Order I CYCADALES

FAMILY I. **CYCADACEAE** Persoon

Shrubs or small trees, with a thick unbranched stem (rarely branched) with a terminal crown of rosette of *leaves,* or stemless with leaves arising from a tuberous rootstock. Leaves palm-like pinnate rarely 2-3 pinnate, leaflets coriaceous or membranous. *Male, Female reproductive* organs on different plants; *males* in one or more terminal cones formed of fleshy flat or peltate scales bearing beneath crowded 1-celled anthers; *females* of flat carpellary leaves (*megasporophylls*) crowded round the apex of the stem (*Cycas*) or cones of flat or thickened peltate scales. *Ovules* large sessile, numerous and erect, in notches on either margin of megasporophyll or solitary. *Seeds* large, drupaceous, external outer coat fleshy, internal crustaceous or bony. Genera 9, 4 of which (*Dioon, Ceratozamia, Microcycas, Zamia*) belong to the Western Hemisphere, and 5 (*Cycas, Macrozamia, Encephalartos, Stangeria, Bowenia*) to the Eastern. The most comprehensive treatment is by R. Pilger 1926 in Gymnospermae in Engler & Prantl Die Naturlichen Pflanzenfamilien ed. 1960. D.D. Pant (1962) has monographed the Cycadales and updated his work in the 2nd 1973 edition.

CYCAS L.

Sp. Pl. 1188, 1753, Mant. 2: 166. 1771

Etymology : The name comes from the Greek *Kykas* used by Theophrastus for this plant.

Evergreen shrubs or trees with simple or rarely forked columnar trunk covered with woody bases of petioles. *Leaves* in terminal crowns linear-oblong, pinnate; leaflets linear, 1-nerved, entire, rolled inward in bud, lower often reduced to spines. *Male* cones terminal, peduncled scales cuneate, imbricate, apex acuminate, anthers ellipsoid, in groups of 3-5. *Megasporophylls* numerous, crowded at the apex of the trunk, densely wooly, appressed into a terminal cone, then spreading, elongate, flattened, dilated above into an entire, crenate, or pectinate blade. *Ovules* 1-5 in notches on either side of stalk of the megasporophyll, distant, alternate or opposite, nearly erect. *Seeds* ellipsoid or globose. Species 20 (Willis 1973). Madag., E & SE AS., Indomal., Austr., Polynes.

Propagation : By seed or basal suckers.

Five species are found in our area. *C. revoluta* Thunb. of Japan is commonly cultivated in Indian gardens.

Key to Cycas

I. Margin of leaflets revolute

A. Small palm-like tree, *c.* 1.8 (3 m). leaflets *c.* 6 to 8 mm wide. Blade of *megasporophyll* pectinate............................ 1. *C. revoluta*

B. Small shrubs *c.* 15 cm high, leaflets narrow 2-4 mm broad. Blade of megasporophyll dentate lobate............................ 2. *C. beddomei*

II. Margin of leaflets flat

Trunk prominently swollen at base. Leaflets not more than 8 mm broad. Blade of *megasporophylls* ovate pectinate along the margin... 3. *C. siamensis*

Trunk not swollen at base. Leaflets usually not more than 12 mm broad.

B. Blade of megasporophylls spinous toothed along the margin...... 4. *C. circinnalis*

Leaflets 12-16 mm broad. Blade of megasporophyll ovate to ovate-lanceolate, densely clothed with long and soft hairs, teeth small..... 5. *C. rumphii*

Leaflets 5-10 mm broad. Blade of megasporophyll broadly orbicular, long acuminate and deeply pectinate................... 6. *C. pectinata*

1. **Cycas revoluta** Thunb. Fl. Japan 22, 1784. Jap. *Tosso, Sotetsu;* Tam. *Madangameswari* — **SAGO SYCAS** (Pl. 1, Fig. 1)

This cycad of Japan and China is the most widely cultivated in gardens in the Indian subcontinent including Sri Lanka. The specific name points to the revolute margin of the pinnae. Commonly the female trees are found introduced in India. A few male trees are to be seen in the Botanical Garden of the Forest Research Institute Dehra Dun where the male cones mature to 30 cm in length in July. The largest known trees 8 m tall are seen in Tyugeji Temple in Tokyo.

Description : A slow growing, palm-like dioecious tree or shrub 1.8 (3 m) producing suckers. Trunk columnar unbranched, rarely forked clothed with old leaf bases. *Leaves* 60-150 cm long, petiole thick quadrangular; leaflets 9-18 cm long, less than 5 mm wide, sub opposite, stiff terminating in a sharp pointed tip, margin revolute. *Male cones* apical cylindrical or ovoid oblong 8-40 cm long, 1.5-4 cm in diameter, shortly peduncled. *Microsporophylls* (stamens) lanceolate-cuneiform 2-3.8 cm long, 11-17 mm wide, truncate, umbonate, woody, covered with pollen sacs beneath. The apical end is sterile, lower wedge shaped is fertile.

Megasporophylls forming an apical crown in a rosette form, densely hairy, stalk longer than blade with 4-6 ovules that are covered with thick down; blade ovate lacinate nearly to the midrib with 12-18 linear curved spinous segments. *Seeds* flattened, glabrous 1.5-3.5 cm long, apex emarginate, bright orange or yellow in colour.

Leafing : New leaves March-April. *Male cones* mature about July.

Distribution : S. Japan, China, Taiwan. Commonly cultivated in gardens in the subcontinent.

Propagation : By seed or basal suckers. Very hardy, resists frost in Dehra Dun and elsewhere in the sub Himalaya.

The pith yields Sago and the fruits can be eaten being rich in protein. The leaves after silvering are made into funeral wreaths (Raizada and Sahni 1960).

2. **Cycas beddomei** Thistleton-Dyer in Trans. Linn. Soc., Ser. 2. 2.11.85 t 17, 1883. Hook.f. FBI: S, 658 1890, Brandis Indian Trees 698, 1906. Raizada & Sahni Ind. For. Rec. NS Bot 5(2) 95. 1960 Pant 1973 vern. *Konda Itha, Perita.*

BEDDOME'S CYCAS
(Pl. 2, Fig. 2)

Etymology : The specific name is to honour Col Beddome a prominent botanist who worked on Indian flora.

Description : A low shrub. *Trunk* short up to *c* 40 cm high. *Bark* brown, exfoliating in rectangular scales exposing a yellow under-surface. *Leaves c* 0.9 m long, *c* 23 cm wide, rachis subquadrangular, petiole *c* 15 cm long, with a few minute spines in upper one-third, base clothed with tufted tomentum; leaflets 10-17.7 cm × *c* 3 mm wide, with a sharp pointed tip, margins strongly revolute. *Male cone c* 30 × 7.5 cm in diameter, very short peduncled, *microsporophyll* (stamen) with an oblong deltoid and a tapering acuminate apex, the lower erect, the upper strongly recurved in the upper half or 2/3rd. *Megasporophylls* 15-21 cm long, blade 7.6 × 2.5 cm, ovate-lanceolate, tapering to a long acuminate point, strongly dentate-lobate. *Ovules* 2-4. *Seeds* globose 3.8 cm in diameter.

Fl. : Nov.-Dec. *Fr. :* March-May.

Distribution : Mostly endemic in **Cuddaph** District, Andhra Pradesh on dry hills and Tamil Nadu.

Specimens examined : ANDHRA PRADESH, Cuddapah Hills, *H.H. Gaide s.n.*

Conservation : Being endemic it is now on the endangered list of flora. Schuster (1932) regards this species as a variety of *C. circinnalis* L emend. Schuster, sub sp *vera* Schuster. According to Pant and Nautiyal (1963) 'its pinnae are clearly distinct in having a revolute margin, a mucilage canal on the abaxial side of the midrib bundle and in their characteristic epidermal and stomatal structure'.

3. **Cycas siamensis** Miq. in Bot Zeit. 4, 21: 334, 1863; FBI 657 Brandis Indian Trees 698, 1906 Pant 39, 1973. Burm - *Mondaing,* Thai : *Prong,* Laos : *Ma Prou Tou.* **SIAM CYCAS** (Pl. 3, Fig. 2)

Description : A small palm-like trees or shrub. Trunk up to 2 m **high,** 30 cm in diameter, 4.5 m tall in Malaya, cylindric, base abruptly swollen. *Leaves* 0.6-0.9 m long, stiff, rusty villous while young; leaflets *c* 70 pairs 7.5-20.5 cm long, 5-7 mm wide, margins flat, mucronate, acuminate, petiole *c* 30 cm long, spiny. *Male strobilus* ovoid oblong. 30 cm long, 6-8 cm in diameter; microsporophylls *c* 19 mm long with a slender terminal point as long as the deltoid clavate limb. *Megasporophyll* ovate-rhomboid, at first densely tawny tomentose, turning glabrous, margin deeply pectinate, lacerate, with 30 processes 2.5-5 cm long. *Ovule* 1 on each side of the stalk at the base of the blade. *Seed* ovoid-oblong, smooth 3.7 cm long.

According to Pant (1973) the megasporophyll margin is spinous dentate in Burmese specimens, almost subulate and entire in those from Cochin-China.

Fl. : March-April.

Distribution : Burma, dry forest of Prome District and of Upper Burma, Thailand, Cochin-China and Yunnan.

Specimens examined : THAILAND, Lampang (N Thailand), *Amnat* 30. Hooker (1890) and Burkill (1933) consider it to be very close to *C. pectinata* and regard it as a geographical race. According to Pant and Nautiyal (1963) the epidermal structure and anatomy of the pinnae in the two species are similar. Schuster (1932) recognizes it as a distinct apecies.

4. **C. circinnalis** Linn.* Emend. Schuster *Cycas circinnalis* (sic) Linn. Sp. Pl. 1188. 1753. Dyer in FBI V: 656 1890 Brandis Ind. Trees. 698 Raizada & Sahni 95, 1960 Pant 34, 1973.
Sanskrit *veraguna;* Hindi *Jangli-madanmast ka phul;* Tel. *Per ita, Kamkshi;* Tam. *Canningay, Madanagama;* Kan. *Mundicalu;* Mal. *Intalappana, Todda Panna;* Uriya-*Oruguna, Odasi-Mari.*

CROZIER CYCAS
(Pl. 1, Fig. 2)

This is the type species of Cycas. the specific name points to the circinate vernation in young foliage.

Description : An evergree palm-like tree 4(-8) m high. Trunk cylindrical rarely forked, with a crown of glossy, fern-like, stiff but gracefully curved, pinnate leaves; trunk covered with compacted woody bases of petioles. *Bark* brown, smooth below, with diamond-shaped scars of fallen leaves. *Leaves* pinnate 1.5-2.5 m long; petiole 0.4-0.6 m long, with short distant spines at right angles to the petiole or slightly deflexed near the base. Leaflets 80-100 pairs, alternate 15-30 cm long, 7-12 mm wide, blunt or acute, margins flat. *Male cone* shortly peduncled *c* 33 cm long, 7.6 cm in diameter, slightly stipitate; microsporophylls 3.7-5 cm long, 1.2-1.9 cm wide, ovate-deltoid, prolonged into an up curved subulate spine, *c* 2.5 cm long, clothed with a brown tomentum externally, glabrous above. *Megasporophylls* brown tomentose, 15-20 cm long, blade 7.6 cm long × 2.5 cm broad, rhomboid, tapering to an acuminate point, margin of megasporophyll crenate or spinous toothed. *Ovules* 6-12. *Seeds* ovoid, pale reddish yellow, 3.8 cm in diameter.

Fl. : Feb.-March. *Fr. :* ripen Aug.-Oct.

Distribution : Western peninsular India, deciduous forests of the W Ghats from sea level to 1070 m/3500′ from Malabar Southwards to hills on

*The valid spelling according to *Index Kewensis* (1895) is *C. circinnalis.* The taxon is commonly spelt as *C. circinalis* L with one 'n'. The latter spelling is by Roxburgh *Fl. Ind* 111, 744. who lists it as a synonym of *C. rumphii* Miq.

the east side as far north as Orissa, Sri Lanka and Maldives near Mala (Fosberg 1957). E tropical Africa, Madagascar, Comora Islands, Sumatra, Java. A form with stunted habit and narrower leaflets is found around Bangalore but is not common.

Specimens examined : TAMIL NADU, Loc. Bellary, 3000′, *E.D.M. Hooper* 11701, in 1883.

C. circinnalis L. var. *orixensis* occurs in the hill forests of the Mals of Puri.

This Cycad (not variety) has a very wide distribution. It has arractive fronds, is more graceful than the commonly introduced *C. revoluta.* It is worthy of cultivation for landscaping around modern buildings where there are limitations of space. It looks graceful in the quadrangles of the Forest Research Institute, Dehra Dun where it was planted about 1976. According to Brandis 1906 it attains an age of 100 years.

5. **Cycas rumphii** Miq. Sc. Phys. Neerl. 45, 1839. FBI 657. Brandis 698. Raizada and Sahni 97. Pant 36. Tam. *Kama, Paiyindu;* Tel. *Rana-guvva;* Mal. *Tuta-ppana;* Kan. *Gudduyicalu;* Oriya *Rosaimaro;* Burm. *Mondaing.*

RUMPHIUS CYCAS
(Pl. 2, Fig. 1)

Etymology : The name commemorates Rumphius who worked on plants of the Asia Pacific region.

Description : An evergreen palm-like tree 3-7.6 m high with a rough simple or often forked trunk not swollen at the base. *Leaves* 0.9-1.8 m long dark green and glossy arising from the top of the trunk; petiole broad at the base, spiny, leaflets 50-60 pairs, 20-38 × 1.2-1.7 cm merging in the upper part of the leaf, linear-acute. Coriaceous, margin flat. *Male strobilus* shortly peduncled 30.4 × 10 cm, ellipsoid-oblong, reddish tomentose, coarsely scented; *microsporophylls* 1.9 cm long, cuneate with a fairly long upcurved not spinescent acumen. *Megasporophylls* c 22.8 × 1.2 cm, blade ovate to ovate lanceolate, densely villous, spinous-toothed above, entire below, abruptly acuminate. Ovules lodged in the notches. *Fruit* globular, c 5 cm across, smooth orange-coloured when ripe. *Seed* ellipsoid 5-7.6 cm × 2.5 cm, orange coloured. 2n = 22.

Fl. : Cold season. *Fr.* : hot season and rainy season. (Sahni 1953 Nicobar expedition).

Distr. : Littoral forests Andamans, Nicobars, Cocos Isls, Sri **Lanka**, S Tennasserim, Malay Peninsula and Archipelago, Molucca, New Guinea and N Australia.

Specimens examined : *K.C. Sahni* 23020 Gt. Nicobar, ANDAMANS, middle Andaman *Kirat Ram* 3806: BURMA *C.E. Parkinson* 14456. Separated from *C. circinnalis* to which it is closely allied by having scales of male inflorescence shortly acuminate, scales of male inflorescence tapering into a long spine in *C. circinnalis.*

Uses : A sago is extracted from the trunk. The fruit is edible when cooked and is eaten in the Andamans. Poisonous when raw. *C. rumphii, C. revoluta* and *C. circinnalis* produce a crown of leaves in the spring and during the monsoon (Pant 1972, p. 48).

6. **Cycas pectinata** Griff. Notul. Pl. Asiat IV, t 360, f.3 10 t, 1854. Thistleton-Dyer in FBI 5: 657. Brandis 698; Raizada and Sahni 96; Tuyama in Hara Fl. E. Him. 9. 1966. *Cycas circinalis* Linn subsp. *vera* var *pectinata* (Griff) Schuster in Engler Pflanzenr. 99, 4(1): 68 1932. vern Nepal *Thakal* N. Bengal *Siapdung* (Lepcha), Assam *Thaljimura* Khasi *Diangsia goda;* Burm *Mondaing, madai.*

NEPAL CYCAS
(Pl. 3, **Fig.** 1)

Etymology : The specific name refers to the pectinate blades of the megasporophylls.

Description : A palm-like evergreen tree 2.4-3 (-6.7) m tall, sometimes with a forking trunk. *Leaves* 1.5-2.1 m long, petiole *c* 44.7 cm long with a few distant spines. Leaflets 17.5-25 cm × 4 to 12 mm, narrow linear, tapering into a minute spine; subfalcate, margin flat-brown, pilose, with deciduous hair, especially on midrib. *Male cone c* 45.5 cm × 15.2 cm, cylindric ovoid, shortly peduncled microsporophylls 3.5-5 cm × 1 to 2.4 cm, deltoid clavate, thickened, abruptly acuminate, acumen *c.* 4 cm long, subulate ascending. *Megasporophylls* 15 to 20 cm long, densely tawny-silky throughout, blade broadly orbicular 7.5 cm wide, long acuminate, acumen 2.5 cm long, margin deeply subulate pectinate, teeth 1-2 cm long, stalk about equal in length to the blade, with 2 or 3 pairs of ovules above the middle. *Seeds c* 3.8 cm long, ovoid, glabrous, orange or yellow orange.

Fl. : Nov.-Dec. *Fr.* : Dec-Jan (Cowan & Cowan 1929).

Distribution : NEPAL in foothills, DARJEELING-SIKKIM, in Sal forests, BIHAR, Sameshwar hills and plains, ASSAM, Goalpara, Kamrup, MEGHALAYA, Khasi Hills, MANIPUR, BANGLADESH (Chittagong), BURMA, pine forests of Martaban and Tenasserim.

In Assam and Meghalaya the tender fleshy shoots are eaten as a vegetable and the seeds are also edible.

Bibliography

Burkill, I.H. : *A Dictionary of the Economic Products of the Malay Peninsula 1.2* London 1933, 1935.

Cowan, A.M. and J.M. Cowan : *The Trees of Northern Bengal.* Calcutta 1929.

Pant, D.D. : *Cycas and the Cycadales.* Indian Universities Press, Allahabad 1973.

Pant, D.D. & Nautiyal, D.D. : Cuticular and epidermal studies of some modern cycadean leaves, sporangia and seed. *Senck Biol. 44*(4): 257-347, 1963.

Pilger, R. (1926) in A Engler & K. Prantl in Die Naturlichen Pflanzenfamilien Gymnospermae ed. 1960 pp 44-82. Duncker & Humblot Berlin.

Schuster, J. *Cycadaceae* in Engler, A. Das Pflanzenreich 4.1 Leipzig.

Willis, J.C. *A Dictionary of the Flowering Plants and Ferns.* ed. 8 (revised by H.K. Airy Shaw) Cambridge University Press 1973.

Order II* **GINKGOALES**

FAMILY 2. **GINKGOACEAE** Engl.

GINKGO FAMILY

Engler in Engler u Prantl, Nat Pflanzenfam. Nachtr II-IV. 19 1897.

A monotypic order and family founded on the genus *Ginkgo.* Stem of trees, tall much branched, leaves numerous. *Male* strobilus catkin-like, stamens in pendulous catkin-like spikes; stamen a short stalk bearing generally two pendent sporangia, dehiscing longitudinally. Ovules in pairs on an axial structure. *Seed* drupe-like, the embryo with 2 dicotyledons embedded in copious endosperm.

Leaves deciduous, an uncommon feature in gymnosperms met with only in *Larix* and *Taxodium* (N. America). It differs from the Coniferae in its fan-shaped leaves, and in the presence of swimming male cells or sperms similar to those found in Cycads. Motile male sperms are ciliated and are also found in Tree ferns. The ovules are fertilized by motile sperms conveyed to them by pollen tubes. This was reported for the first time by Hirase 1895-6 from Japan. According to Dallimore and Jackson (1966), 'Fertilization is completed in autumn, but the development of the embryo is not finished until after the seeds have fallen to the ground'.

Ginkgo biloba is the only living representative of an ancient order Ginkgoales of conifer-like trees. It is the only species of the one genus of the family and of the order. It is sacred to the Buddhists and has been cultivated for centuries in China and Japan in temple grounds. According to Alvin (1981) 'there is now good evidence for believing that it occurs in a truly wild state at the borders of Chekiang and Anhwei provinces in eastern China'. In India it is seen in cultivation at Dehra Dun, Mussoorie, Shimla, Ootacamund and elsewhere in moist cooler climates. It is able to withstand smoke and is tolerant of pollution and low water supply. It is therefore useful for industrial areas in Himachal Pradesh, Kashmir and other Himalayan areas and hills above 800 m. Ginkgoes present a glorious sight in autumn when their leaves turn a golden yellow. They are planted in avenues and parks particularly in the U.S.A. Some magnificent trees are growing in the arboreta of the Forest Research Institute, Dehra Dun.

This order flourished during the Mesozoic era, especially during the Jurassic period i.e. about 150 million years ago and was contemporaneous with the ferocious-looking giant sized reptiles the dinosaurs. In the geological past it was world-wide in distribution and today only one species is in existence. Because of its interesting history this living fossil is in demand in Natural History establishments. It is called the Maiden-hair Tree because its foliage resembles *Adiantum* or maiden-hair fern.

Ginkgo Linn.
Linn. Mant. Pl. 2, 313, 1771

1. **Ginkgo biloba** Linn. Mant. Pl. 2 313, 1771; Thunb. Fl. Jap 358 1784; Pilger in Engler u Prantl, Pflanzenfam ed, 2, 13:99 f. 96-101, 1926. Raizada and Sahni 97 1960 Hara, Stearn & Williams Enum. Fl. Pl. Nepal 1:6, 1978. *Salisburia adiantifolia* Smith in Trans. Linn. Soc. London 3:330 1797 *Salisburya biloba* Hoffmagg. Verz. Pflanzenkult 109. 1824. vern. Yin-kuo (China).

MAIDEN HAIR TREE
(Pl. 4, Fig. 1)

Etymology : The name *Ginkgo* is derived from the Chinese 'Yin-kuo (Silver-fruit). The Japanese pronunciation is 'ginko'. Some authorities have adopted the spelling 'Ginko'.

Description : A deciduous tree, diocious, upto 40m/130′, 5-6 m in girth, mostly up to 15m tall in India where it was introduced about 70 years ago. *Bark* greyish, furrowed on old trees. *Leaves* fan-shaped, glossy green, 5-7.5 cm, on young vigorous trees 15-20 cm across, resembling an enlarged pinnule of maiden-hair fern, upper margin irregularly toothed, divided by a deep incision which cuts them into two lobes. There is no midrib, but there are numerous parallel branching veins; stomata mostly on the lower surface. *Strobili* Male and Female usually on different trees (when they occur on the same tree it is owing to shoots of a tree of one sex having been grafted on one of the opposite sex). The male strobili grow as catkins, the stalked females spring from the axils of scale-like bracts and are followed by fruit in the form of false drupes with a fleshy foul smelling pericarp.

Drupe : plum-like *c* 3 cm in length, orange coloured.

Fl. : Spring. *Fr.* autumn. *Lf.* spring. *Lf.* Shedding autumn when they turn a golden yellow. From Herbarium sheet records. *Fl.* male 20.4 1966 *K.M. Vaid* s.n., Fr. 27.6.1965 *K.M. Vaid* s.n. Loc. Dehra Dun.

Distribution : Wild in E China, on the borders of Chekiang and Anhwei provinces. Cultivated in cooler climates world wide. According to Lawrence 1960, it is a relic of once wide global distribution.

Propagation : By seed or by cuttings of softwood, or by grafting in order to eliminate female specimens which are evil smelling in fruiting condition.

Conservation : This most interesting 'Living Fossil' is suggested for ceremonial plantings in cooler climates as it is long lived. It is reputed to live for over 1000 years in the grounds of temples and palaces in Japan and China where it has survived through the conservation efforts of Buddhist monks.

Pollution control : Tolerant to smoke and pollution. In the Indian subcontinent it would flourish in the Himalaya, adjacent hills and hill ranges of Peninsular India, Sri Lanka and Burma.

Uses : It is used in China and Japan for making chess boards and Chess men. The seeds are eaten roasted at feasts aid digestion, and diminish the effects of drinking wine. (Dallimore & Jackson. 1966).

Moule 1944 and Thommen 1949 consider that etymologically the name *Ginkgo* is an orthographic error and that the correct translation of the Chinese characters used to designate this tree should be "Ginkyo". The error originated with Kaempfer (of Dutch E India Company) in 1712 and was introduced into botanical nomenclature by Linnaeus in 1771. Some botanists accept the views of Moule and of Thommen and have rejected the name Ginkgo as "an unintentional orthographic error" as allowed by the Rules, and have replaced it with Ginkyo (Cf. Pulle 1946, Widder 1948).

Kaempfer according to Dallimore Jackson (*loc cit*) discovered the tree in 1660 and published a description in 1712 with illustrations of foliage and fruit. It was first planted in Europe about 1727 in the Botanical Garden Utrecht.

Till the late 19th century *Ginkgo* was included in the Coniferae. Eichler 1883 included it in the Taxeae alongwith *Taxus, Torreya & Cephalotaxus.* Soon after the discovery of motile ciliated sperms in Ginkgo (Hirase 1896). Engler 1897 created a new class (order) the Ginkgoales, for *Ginkgo* and its fossil members.

Bibliography

Alvin, K.A. in *The Oxford Encyclopedia of Trees of the World* (Ed. Bayard Hora) O U P. Oxford 1981.

Dallimore W, and A.B. Jackson. *A Handbook of Coniferae* and *Ginkgoaceae* rev. Harrison. London 1966.

Eichler, W.A. *Syllabus der Vorlesungen* ed 3. 1883.

Engler A. (in Engler & Prantl), *Die naturlichen Pflanzenfamilien* Nachtrag zu Teilin II-IV 1897.

Hirase, S. Etude sur le fecondation et l' embryogenic du *Ginkgo biloba J. Coll.* Sci. *imp. univ.* Tokyo. 12: 103-149, 1896.

Lawrence, G.H.M. *Taxonomy of Vascular Plants* ed. 5. Macmillan Coy. New York. 1960.

Moule, A.C. The name *Ginkgo* Journ. Roy. Hort. Soc. 69: 166 1944.

Pulle, A.A. Over de *Ginkgo* alias *Ginkyo.* Jaarboek Nederl. Dendrol. ver. pp 25-35, 1940-1946.

Thommen, E. Neues zur Schreibung des Namens *Ginkgo.* ver. Nat. Gas Basel 66: 77-103. 1949.

Widder, F. Die Rechtschreibung des Namens *Ginkgo.* Phyton 1: 47-52, 1948.

Order III **TAXALES** B. Sahni **TAXADS**

Birbal Sahni (1920) separated the Taxaceae from the Coniferales and aised this family to a new order Taxales within the Division Gymnospermae. He was no doubt struck by the fact that the female eproductive organ in the Yew family is not cone like but is characterized by a fleshy red aril enclosing a bony seed. Florin (1948) up held Sahni's separation of the Taxales as a new order equal in rank with the Ginkgoales and Coniferales. Takhtajan (1953) also supported Sahni's classification and in addition retained his nomenclature. Pilger (1916) elevated Taxaceae to the rank of an order a view he rejected later in 1926 in Taxaceae in Engler and Prantl Die naturlichen Pflanzenfamilien ed 2. Bd. 1: 199-211, 1926. Harrison (1966, p. 1) in his revision of Dallimore & Jackson's Handbook of Coniferae and Ginkgoaceae has said "There is an increasing body of opinion in favour of placing the family *Taxaceae* in a separate order the Taxales". This view is also shared by Airy Shaw 1973, p 1132 in his revision of Willis *Dict. of Fl. Plants and Ferns.*

Conforming to modern trends the order Taxales is restricted to the single family Taxaceae which has been removed from the Coniferales and placed in a separate order within the Gymnospermae in the present work. This arrangement is also followed by Harrison (loc. cit.). In our area Taxales is represented by two genera *Taxus* and *Amentotaxus.*

Taxus has been kept in a Tribe Taxeae by Melchior et Werdermann in Engler's Syllab. Pflanzenfam. ed. 12.1.341, 1954.

Amentotaxus in Tribe Amentotaxeae by Cheng et C.D. Chu in Gymnospermae, Fl. Rep. SINICAE. Tomus 7, 437, 1978.

Family 3. **TAXACEAE** S.F. Grey Nat. Arr, Brit. Pl. 222. 226. 1821
YEW FAMILY TAXADS

Confined to Northern temperate forests to Himalaya, adjacent hills of N.E. India, Philippines, Celebes and Mexico. The type species is *Taxus baccata* L., the common yew. Evergreen trees or shrubs. *Leaves* persistent spirally arranged, rarely opposite (*Amentotaxus*), mostly 2-ranked on horizontal shoots, usually flat needle-like, linear without a resin canal or scale like. *Male* and *female strobili* usually on different trees (rarely on the same tree). Male strobili catkin-like, their sporophylls radially symmetrical in some genera dorsiventral in others. *Pollen* grains wingless. Female strobilus not in the form of a cone, axillary, with a single, *ovule* or seed, erect and bony, not borne on a scale, but partly or wholly surrounded by a fleshy highly coloured cup (aril) when ripe.

It was placed by several authorities in the order Coniferales till recent times. The Yews (*Taxus*), *Amentotaxus* and others of the family Taxaceae lack the typical seed bearing cone of typical conifers or 'cone bearers' and do not have resin canals in the wood and leaves. In view of these characters, coupled with palaeobotanical evidence, this family is excluded from the Coniferales and placed in a separate order Taxales. When not in fruiting condition, *Taxus* resembles the foliage of Fir (*Abies*) and Hemlock (*Tsuga*), but is at once distinguished by field characters of the latter two, characterized by a tall trunk (Fir), tall and massive trunk (Hemlock). The underside of the leaves of Taxaceae is yellow-green without prominent stomatal lines which are seen in Fir and Hemlock.

Genera : *Taxus, Pseudotaxus* Cheng (Syn Nothotaxus Florin), *Torreya, Austrotaxus* and *Amentotaxus.* The first and the last named occur in our area.

Key to *Genera

Leaves linear, opposite. Male strobili on spike-like racemes, usually in groups of 2-4. Ovule solitary in the axil of a bract. 'Fruit' drupe-like, reddish-yellow aril open at the apex .... AMENTOTAXUS

Leaves linear, more or less spreading 2-ranked. Male strobili stalked with the sporophylls borne terminally in sub-globose heads, surrounded by scales at the base. Ovule solitary; aril cup-like red .... TAXUS

1. **TAXUS** Linnaeus. YEW
Linn. Gen. Pl. 312 1737 Nr 756.

Evergreen trees or shrubs with linear 2-ranked *leaves.* Winter buds with obtuse scales. *Male* and *female strobili* usually on different trees (rarely on the same tree). Male strobili in stalked, sub-globose heads; anther scales peltate. *Seed* erect with a bony shell, borne in a fleshy cup or aril normally

*Covering India/adjacent countries.

red, ripening in the first year. Species 10 (regarded by some as varieties of one species). Yews are widely distributed throughout the N Hemisphere in the temperate zone, with one species, the Chinese Yew (*T. celebica*) on the equator on the Indonesian Island of Celebes (Sulawesi), one in New Caledonia. Only *Taxus baccata* occurs in our area.

Taxus baccata L. Sp. Pl. : 1040 (1753) subsp. **wallichiana** (Zucc.) Pilger in Engler, Pfl. reich IV-5 Ht. 18: 112 1903. FBI: 5, 684 1890. Brandis Indian Trees 696, 1906. Raizada & Sahni 103, 1960 Dallimore & Jackson Ed 4. 587, 1966. Hara, Stearn & Williams 1:6 1978.

COMMON YEW
(Pl. 4, Fig. 2)

W. Him., *Thuner,* Kash. *Birmi, Postil;* Beng *Bhirmil;* Lephca *Cheongbu;* Nepal *Tcheiragulab;* Khasi *Barmi, Birmi,* Burm. *Kyauk-tinye.*

Description : A slow growing small evergreen tree up to 6 m and 1.5-1.8 m girth with a fluted stem. *Bark* thin reddish brown. *Branchlets* spreading, not whorled. *Leaves* 2.5-3.8 cm long, linear, flattened, arranged in 2 opposite vertical rows, dark glossy green above, paler beneath. *Male & female strobili* usually on different trees (rarely on the same tree). Male strobili in catkins which are sub-globose and solitary in the leaf-axils, stamens *c* 10, pollen sacs 5-9, globose, arranged around the filament beneath the tip of the stamen. Female strobili solitary, axillary, each consisting of a few imbricate scales around an erect ovule, which is surrounded at the base by a membranous cup shaped disk. In *fruit* the disk enlarges, becomes succulent and bright red, 8 mm long, and surrounds the olive green seed.

Distribution : All along the Himalaya, Meghalaya, Naga Hills, Manipur and in Burma mostly between 2300-3400 m in the evergreen and coniferous forests. Slow growing and long lived, up to 2000 years. Also in Afghanistan and SW China.

Lf. shedding May - June. *Fl. :* March - May. *Fr. :* Seeds ripen Sept. to Nov. of the same year.

Specimens examined : ARUNACHAL PRADESH *K.C. Sahni* 5091, Peri La Kameng Dt. 2,700 m. *Birds* swallow the aril and thus distribute the seed. NEPAL* *Polunin, Sykes & Williams* 1873 (BM), *Stainton* 4496 (BM)

Poisonous effects : All parts of the Common Yew, except the scarlet aril are highly poisonous. According to Hora 1981 'the seed contains poison, children should be discouraged from eating the tempting red fruit. Yew poisoning, resulting in gastroenteritis; heart and respiratory failure is extermely serious and fatal results both in humans and animals are well documented. There is sudden death within five minutes'.

Propagation : By seed and for cultivars by cuttings, grafting (on stocks of *Taxus baccata*) or by layer.

Uses : It makes an excellent hedge over 6 m tall, as such it is popular in Europe where it also is a favourite for topiary work. In Britain it was the traditional material for making bows and is still in use for archery sports. (Hora loc. cit.). In Ladakh the bark is said to be used in place of tea. It is used locally for ploughs and axils of carts. (Raizada & Sahni 1960).

2. **Amentotaxus** Pilger in Bot. Jahrb 54: 41, 1916: Taxaceae

Etymology : The name is derived from the Latin *amentum,* catkin referring to the pendulous male inflorescence and *taxus.* The genus comprises 1-4 species confined to Arunachal Pradesh, Central and S China including Hongkong, Taiwan and Vietnam. It was formerly placed in the family Cephalotaxaceae.

Evergreen shrubs to small trees. *Branchlets* opposite. *Leaves* decussate, sessile or subsessile, linear-lanceolate, apex acute, strongly keeled beneath which is the midrib, leathery, dark green, bordered on each side by a glaucous stomatic band. The *Male/Female 'cones'* are borne on separate plants. Male strobili in spike-like racemes in groups of 2-4 (rarely 1 to 5), pendent, the strobili composed of clusters of stamens and peltate scales. *Female strobilus* solitary, stalked, with small imbricate scales, in the axil of a lateral bract. *Fruit* drupe-like, the reddish-yellow aril open at the apex, scales persistent at its base.

---

* Not examined

The type species in *A. argotaenia* (Hance) Pilger (basionym *Cephalotaxus argotaenia* Hance), a shrub native to Kwangtung and Lanton Island, Hongkong. Pilger 1916 considered this genus monotypic at that time. Later Li 1952, and Ferguson 1985 were to identify another two from China, Taiwan and one from Arunachal Pradesh.

The one species in our area viz *Amentotaxus assamica* D.K. Ferguson was collected by F. Kingdon Ward No 8026 (K) from 'Delei Valley Chibaon Assam' this locality is in Lohit District, Arunachal Pradesh. This taxon is characteristic of temperate rain forests, at *c* 8000-9000′. This taxon was earlier determined by D.R. Hunt. in 1972 as A. cf. *argotaenia* (Hance) Pilger, according to Ferguson (1935), p. 115. A second specimen J.L. Lister s.n from Dafla Hills was determined by Li 1952 as *A. cathayensis* Li. Ferguson 1985 has identified these two specimens as a new species, *A. assamica* Ferguson on the basis of critical herbarium study and detailed investigation of their leaf anatomy. He has pointed out (p 115) 'These 2 specimens are not only linked by their similar provenance, but in the total absence in the leaf blade of sclereids, whose size and distribution has proved to be the most valuable character in distinguishing the species. All other taxa have such sclereids'. A detailed description is as under, based on Fergusan 1985 amd *Haridasan* No. 2 856 22.3.1986 in DD.

1. **Amentotaxus assamica** D.K. Ferguson K.B. 40(1): 115-119, 1985. (Pl. 5, Fig. 1)

Description : Tree up to 20m, trunk sometimes divided into two low down. *Bark* whitish grey, hard. *Branches* whorled, lower ones often retained, sweeping down, ultimate branchlets opposite, arising at an acute angle, glabrous, square in cross section. *Buds* ovoid, 8.5-17 mm long, 4-8 mm broad, square in cross-section, bud-scales decussate, caducous, outer ones triangular, inner oblong, keeled, mucronate, mucro 0.5-0.8 mm long. *Leaves* opposite, shortly (1-4 mm) petiolate, with a life span of 3 + years, linear-lanceolate, falcate or somewhat S-shaped, (20-) 70-115 (-150) mm long, (3.8-) 7-12 (-12.3) mm broad, gradually tapering to an acute apex, base cuneate, margin entire, sometimes slightly revolute, one-nerved, bifacial, adaxially smooth, sometimes longitudinally striate in dried state with raised mid-rib abaxially divided into 5 zones central, costal, zone 0.9-1.8 mm wide flanked on either side by an initially white to silver-grey stomatal band (0.65-) 1.5-3 mm wide, marginal band (0.4-) 1.1-2.1 mm wide. *Staminate* racemes borne terminally, 4 per bud, 40-55 mm long at maturity, bearing mostly paired spheroidal, ovoid or ellipsoidal fascicules of stamens up to 6 mm long. Stamens shortly stipitate, peltate, 1-2 mm diam. bearing 2-4 pollen sacs on their lower surface. Pollen inaperturate, spheroidal to ovoid, 28-38 μm diam., exine ± 1 μm thick, scabrate. ♀ *Strobili* axillary, solitary generally one in each node; Scales 8, upper 2.5-5 × 2.5-5 mm, keeled. *'Fruit'* 2-3.5 × 1.5-2.5 cm, oblong-obovoid, shortly beaked at tip, shining, green, turning yellow and later purple. Stalk 1.5-2.5 cm angular *seed* one. 'INDIA'. Assam : Delei Valley, Chibaon, 6 April 1928, *F. Kingdon Ward* 8026 (holotype K); Dafla Hills, 1874. *J.L. Lister* s.n. (A)' *vide* Ferguson loc. cit. These localities along with Dafla Hills are now in Arunachal Pradesh, Lohit District. Endemic, as specimens outside this state have to date not been sighted or collected. Recently collected by Haridasan from Delei Valley, from Mithumna-Mailing 1800 m Lohit District, Arunachal Pradesh (Personal discussion followed by a specimen) See Haridasan 1988.

Specimens examined : ♀ 1443, 18.9.84 not seen, Haridasan sketch examined. *Haridasan* 2856, 22.3.86, male DD:

Lf. : Evergreen : *Fl. :* March-April. *Fr. :* August to Oct.

Distribution : In localities mentioned above. According to Ferguson op. cit. p 119, 'The species is characteristic of temperate rain forest, occuring with *Magnolia, Quercus, Rhododendron*' etc. on precipitous north-facing slopes. Kingdon Ward noted that mature specimens are moss-draped. The branches are similarly moss-draped in J.L. Lister s.n. Altitude is not recorded in Kingdon Ward and Lister specimens. In my opinion it should be *ca* 8000-9000′. The moss is *Barbella spiculata* (Mitt) Broth. (*Meteoriaceae*), which is a light-demanding species typical of forest margins and clearings. (vide Ferguson loc. cit.). Haridasan says 'in dense temperate broad-leaved forests, associated with *Rhododendron* spp., *Castanopsis, Exbucklandia populnea* nearly gregarious on hill slopes'.

A Rare/Endangered Taxon : Endemic and certainly an extremely rare Taxaceae as evidenced by the fact that inspite of several surveys by Sahni, 1969, 1970, 1977 and 1973 and by botanists of the Botanical Survey of

India to Arunachal Pradesh a third specimen of this taxon has only recently been reported to the author by Haridasan (1988). It was first collected in 1874 by Lister and after over half a century later in 1928 by Kingdon Ward.

Conservation : There is urgent need to identity and demarcate certain areas in Arunachal which are rich in rare taxa e.g. *Sapria, Cephalotaxus, Amentotaxus,* etc as nature reserves, Sancturies, National Parks or Biosphere Reserves. In any case *A. assamica* muzt be reared urgently at Lloyd Botanic Garden Darjeeling and Experimental Garden of the Botanical Survey of India at Shillong. It is certainly vulnerable to extinction in view of developmental activities like road building and the practice of shifting cultivation prevalent in the hills of Eastern India and in the Eastern Himalaya.

Bibliography

Cheng Wan-chun, Fu Li-Kuo et Chu Cheng-de. Gymnospermae in *Flora Reipublicae Popularis Sinicae,* Science Publication, Beijing, 544 pp, 1978.

Ferguson, D.F. A new species of *Amentotaxus* (*Taxaceae*) from north eastern India. *Kew Bull. 40(1): 115-119, 1985.*

Florin, R. Die Koniferen des Oberkarbons und des unteren Perms Paleontographica 85: 1-654, 1938-44.

Haridasan, K. (1988). *Amentotaxus (Taxaceae).* A rare Gymnosperm from Arunachal Pradesh. *Indian For.* 114(12) 868-870.

Hora, F.B. *The Oxford Encyclopedia of Trees of the World,* Oxford University Press, 1981.

Hunt, D.R. *Amentotaxus* from Assam. *Kew Bull.* 22(3), 402, 1968.

Li. J. Arnold. *Arbor. 33(2) 192-8, 1952.*

Melchior, H. & Werdermann, E. in Englers Syllabus der Pflanzenfam. ed. 12, 1.34 1954.

Pilger R. Die Taxales Mitt. Deutsch. Dendr. Ges. 25: 1-28. 1916. Taxaceae in Engler and Prantl. Die naturlichen Pflanzenfamilien ed. 2. Bd. 13: 199-211, 1926.

Raizada, M.B. & K.C. Sahni. 'Living Indian Gymnosperms' pt I (Cycadales, Ginkgoales Coniferales). *Ind. For. Rec.* 5(2): 1-10, 73-150, 1960.

Sahni, B. On certain archaic features in the seed of *Taxus baccata* with remarks on the antiquity of the Taxineae. *Ann. Bot.* 34(133): 117-133, 1920.

Sahni K.C. Contribution to the Flora of Kameng and Subansiri Districts, NEFA. *Indian For.* 95(5): 330-52, 1969.

—— 'Protection of Rare and Endangered Plants in the Indian Flora' *Proc. IUCN* 11th *Technical Meeting.* Morges, Switzerland 11: 95-102, 1970.

—— Tour Report, Kameng Forest Division, Arunachal Pradesh, April 1977 (unpublished).

—— Tour Report, Kameng Forest Division, Arunachal Pradesh, 1973 (unpublished).

Takhtajan, A.L. Phylogenetic principles of the system of higher plants. *Bot. Rev.* 19: 1-45, 1953.

Order IV. **CONIFERALES**

The families included here are the *Cephalotaxaceae, Podocarpaceae, Araucariaceae, Pinaceae, Taxodiaceae* and *Cupressaceae.* Dallimore and Jackson (1966, p.1) rev. Harrison Handb. Conif. Ginkgo have pointed out 'There are conflicting views on the limits of the families and genera to be included in the Coniferales'. The Taxaceae formerly included under the coniferales has according to modern concepts been removed from the Coniferales and raised to the status of an order Taxales. This has been discussed at length under the Taxales. The classification of the genera and families comprising the Coniferales has not yet become stabilized. The inclusion of Cephalotaxaceae and Podocarpaceae is debatable as they are not typical conebearers. The families *Cephalotaxaceae, Podocarpaceae* (and *Taxaceae*) because of the absence of typical (female) cones are conveniently referred to as 'Taxads' to distinguish them from the remaining four families which have recognizable cones and are correspondingly known as pinoids. Pinoids are adapted to wind dispersal of their seeds e.g. winged seeds of *Pinus.* The exception is *Juniperus.* Taxads are adapted to dispersal by animals and birds. From the point of view of economic importance conifers are of major importance in our area (Deodar, Pine) and in the world.

The families of coniferales are treated in their systematic sequence as enumerated on pp. 22 starting with Cephalotaxaceae and ending with Cupressaceae.

Family 4. CEPHALOTAXACEAE Neger Nadelh 23.30. 1907
PLUM-YEW FAMILY

Small evergreen yew-like trees or shrubs. Leaves alternate, spirally inserted, are spreading on vertical shoots, distichous on lateral shoots, narrow linear, flat, pointed, shining green above. Similar to those of *Taxus,* easily distinguished by the presence of large resin canal on the lower side between the midrib and epidermis. Midrib prominent. *Flowers,* the sexes are typically on separate plants, male forming globose heads in the axils of membranous bracts. Stamens 7-12, with short filaments and usually 3 sporangia (pollen-sacs). Female flowers in pedunculate cones at the base of branchlets, each comprising a few pairs of scales, and each bearing 2 ovules. Usually only one fertilized ovule in the cone matures to form a drupe-like seed. Only genus : *Cephalotaxus* found from E. Himalaya to Japan.

1. **Cephalotaxus** Siebold & Zuccarini ex Endlicher Sieb. et. Zucc. ex Endl. Gen. Suppl. 2.27.1842. **COW'S TAIL PINES, PLUM YEW**

Evergreen trees or shrubs. Branches opposite or in whorls. Young branchlets green, prominently grooved and marked with minute white stomatic dots. Buds with numerous overlapping scales. *Leaves* spirally arranged and spreading or vertical shoots; on lateral shoots arranged in two opposite ranks. Leaves with a conspicuous midrib. Male and female strobili on different trees or rarely on the same tree. Male strobili in globose heads. Female strobili few, stalked, composed of opposite pairs of cup-shaped bracts with 2 ovules at base of each. Usually only one of these develops into an olive-like 'fruit' (seed), consisting of an inner, woody, acron-shaped shell. There are 2 species in our area which occur in N.E. India and Burma. The genus has 4-seven species. Species not in area are in China (*C. fortuni*), Japan and China *(C. harringtonia)* and Korea.

Key to Cephalotaxus

Leaves narrowed towards the base, pale green beneath................ 1. *C. mannii*

Leaves with a comparatively broad rounded base, whitish beneath...... 2. *C. griffithii*

C. mannii Hk.f. and *C. griffithii* Hk.f. are small evergreen trees found in the evergreen hill forests of the Khasi and Naga Hills as well as the evergreen forests of the Eastern Himalaya, the latter extending to hills of Upper Burma. Their timber does not appear to be used. The former species has been introduced into Europe where it does well particularly in the Rhine Valley.

1. **Cephalotaxus mannii** Hk.f. in Hook. Icon. Pl. t. 1523 m 1886. et FBI: 5: 647 1888. Pilger in Engler Pflanzenr. 18 Heft 4(5): 103, 1903. Brandis Ind. Trees 697 1906. Raizada & Sahni 107, 1960. Dallimore and Jackson Handb. Conif. Ginkgo Ed. 4, 152, 1966. **KHASI PLUM YEW** (Pl. 6, Fig. 2)

Description : A small tree. *Bark* cinnamon-brown. *Leaves* 2.5-5 cm long by 0.25 cm -0.38 cm, linear, sub-falcate, gradually narrowed from the base, apiculate, under-surface white; anthers about 12, mucronate. *Female cones* 3-4, on a common peduncle; bracts adnate, ovate, obtuse. *Seeds* 3-4 on a common stalk, each 3-3.8 cm long, obovoid, with a short apical point, narrowed at the base.

Distr. : Khasi, Jaintia, and Naga hills from 1370-2590 m. Upper Burma hill forests.

Flowering/fruiting : Specimens with 'flowers' and seeds in Herbarium Dehra Dun collected by G. Mann are dated August and November, 1885. Specimens examined : Meghalaya : Muplang 5-6000′ Khasi Hills, Aug. 1885 *G. Mann.*

2. **Cephalotaxus griffithi** Hook.f. FBI V: 648 1888. **PLUM YEW**
Raizada & Sahni 107, 1960. Dallimore and Jackson
Jackson Handb. Conif. Ed. 4, 149, 1966. (Pl. 6, Fig. 1)
Burm. *Tinya*

Description : A middle-sized tree. *Bark* brownish smooth and polished. *Leaves* 2.5-5 cm long by 0.25.0.38 cm falcate; sub-acute; base inequilateral rounded, whitish beneath. *Seeds* c 3.7 cm long, borne 3-4 together sometimes longer, with a short apical point.

Distribution : Kameng District, Arunachal Pradesh, Naga Hills, Manipur and Mishmi Hills 1520-1820 m and in Burma.

Specimens examined : *K.C. Sahni & H.B. Naithani* 903 Jameri, Kameng District, Arunachal Pradesh. *K.C. Sahni 601,* Ramda, Kameng District, Arunachal Pradesh. BURMA 5600′ *J.W. Oliver* s.n.; Kawhip 8000′, 9000′ *U. Pokhant* 2437; 9000′ *Mg Gala* 5977.

Propagation of allied species which have been tried is by seed or cuttings.

Conservation : These two species are very scarce in the E Himalaya, Naga Hills, etc. They occur in areas of shifting cultivation and are recommended to be put on the Endangered list of flora.

Family 5 **PODOCARPACEAE** Endl. Endl. syn. Conif. 203. 1847.
YELLOW WOODS FAMILY PODOCARP FAMILY

Evergreen trees or shrubs with elliptic-lanceolate, linear-lanceolate or elliptic *leaves* which are usually spirally arranged but sometimes twc ranked as in *Abies,* varying in size from scales to nearly 30 cm long. Young leaves often rose or red in colour. *Male/Female* strobili usually on different trees (dioecious). Male strobili cylindrical, dense, catkin-like, axillary or terminal. Stamens numerous with short filaments; anther cells two, pollen grains winged. Female flowers stalked or sessile, axillary or terminal with usually 2-4 scales one or two of which bear in their axils a fertile scale folded over and united to an inverted ovule; usually one matures, the sterile scale often fusing with the upper part of stalk, and developing a swollen fleshy, brightly coloured edible receptacle, upon which seed is borne. *Seed* round or ovoid, with an outer fleshy or skinny seed coat and an inner woody shell. *Podocarpus* is the only genus in our area. Mostly confined to S. hemisphere extending N to Japan.

1. **Podocarpus** L′ Herit ex Pers. Syn Pl. 2: 580, 1807. **YELLOW WOODS**

Etymology : Derived from the Greek *pous,* foot, and karpos fruit in reference to the fleshy fruit stalk of many species.

Evergreen trees and shrubs commonly called the podocarps or yellow woods mostly confined to the warm temperate and sub-tropical regions of the southern hemisphere extending N to Himalaya and Japan. Branches whorled irregular. Leaves very variable, dense or distant on the branches, spirally arranged, rarely opposite or sub-opposite; varying in size from mere scale-like bodies to a length of 23-30 cm and a width of 5 cm. *Strobili* Male and female usually on different trees. Sometimes monoecious. Male strobili axillary or rarely terminal forming dense narrow cylindrical catkins. Female strobili stalked or sessile, axillary or terminal with usually 2-4 scales; one or two of which bear in their axils a fertile scale folded over and united to an inverted *ovule;* usually one only matures, the sterile scale often fusing with the upper end of the stalk and developing a swollen, fleshy, brightly coloured, edible receptacle on which the drupe or nut-like *seed* is borne. The outer seed is skin-like or fleshy, the inner woody. The genus is unique in having the only known coniferous parasite *P. ustus* which parasitizes *Dacrydium taxoides,* also belonging to the family Podocarpaceae (Hor 1981). The genus includes *c* 100 species.

Key to Podocarpus

Leaves not more than 2.5 cm long .... 1. *P. imbricatus*

Leaves much longer, 10 cm-17.8 cm long.

Leaves linear .... 2. *P. neriifolius*

Leaves elliptic .... 3. *P. wallichianus*

1. **Podocarpus imbricatus** Bl. Enum. Pl. Jav. 89 1827 Pilger in Engler Pflreich IV-5 56, 1903. Brandis Ind. Trees 696, 1906. Raizada & Sahni 104, 1960. Dallimore & Jackson Ed 4, 530 1966. Gymnospermae Fl. Rep. China p 401, 1978. syn *P. cupressina* R. Brown. (Pl. 7, Fig. 1)

Description : Tall tree up to 24 m or more in height with a girth of 3-4.5 m. *Branches* pendulous, spreading, whip-like *leaves* of two kinds, those in young vigorous trees 0.6-1.27 cm long, flat, soft, arranged in two, more or less opposite ranks, those on older trees very short and closely overlapping like those of *Cupressus* (and hence the older name *P. cupressina*), the points being incurved like miniature leaves of *Cryptomeria japonica.* The seed bearing branchlets have small appressed leaves. *Female* strobili almost enclosed by the upper most leaves. *Seed* solitary, borne on thickened fleshy receptacle. Seed and stalk red.

Distribution : Upper Burma (Myitkyina 800-1800 m), China, N Borneo, Java, New Guinea and Philippines, where it is an important timber tree.

Specimens examined : Burma : S Shan States, Myinkyada, Large tree girth 8′, *F.A. Reich* 5846.

Fl, Fr. From herbarium sheet records in D.D. it appears to flower in Burma in June-July and fruits August-Sept.

2. **P. neriifolius** D. Don in Lamb. Desc. Gen. Pinus 2: 21, 1824. Hook.f. FBI 5: 649, 1888. Pilger in Engler Pflanzener. 18 Heft, 4(5): 80, 1903. Brandis Indian Trees 695 1906. Raizada & Sahni 105, 1960 Dallimore and Jackson Handb. Conif. Ginkgo. ed. 4. 538, 1966; Hara, Fl. East. Himalaya syn. *Podocarpus discolor,* Bl. Rumphia 3: 213, 1847. **OLEANDER PODOCARPUS** (Pl. 6, Fig. 3)
Hindi *Halis;* Nep. *Gunsi;* Burm.-*Thitmin-po.*

Description : A large evergreen glabrous tree 12-18 m high with whorled branches. *Bark* greyish brown or light cinnamon brown, thin, peeling off in long papery flakes, 0.4 cm thick, reddish and finely fibrous inside. *Leaves* nearly sessile mostly 10-17.8 cm by 0.8-1.48 cm, on old specimens 12.7-25 cm long and broader in young specimens, mostly 1.27-1.9 cm wide, linear, thick leathery, midrib prominent on both surfaces, apex long. *Strobilus* in clusters or spicate. Peduncle of the female solitary, axillary, single flowered c 1.27 cm long. Receptacle of the fruit oblong, fleshy, enlarging in breadth, with an oblong depression at the top and variously lobed on each side from pale-yellow green turning orange-red, at length deep purple. *Seeds* solitary, globose, 1.5-1.77 cm by 0.5-1.27 cm bluish-black when ripe.

Distribution : It is typically an inhabitant of the evergreen climax forests of Eastern India and Burma. It is found in the E Himalaya up to 900 m in Sikkim, Bhutan, Arunachal Pradesh, Burma, Andamans, Malaya Peninsula extending to S.W. China and Sunda Islands.

Specimens examined : ARUNACHAL PRADESH, K.C. Sahni 559, near Ramda, Kameng District. *K.C. Sahni & Naithani* 923 above Jameri, Kameng Dist. ANDAMANS, Baratang, *C.E. Parkinson* 421.

Fl. : rainy season. *Fr. :* Cold season.

A variety known as var *brevifolius* Stapf has been described as being found in E Himalaya and Khasi Hills.

Uses : The species is used for general carpentry and for making oars, spars and masts. The timber is of high class and deserves to be widely grown.

3. **Podocarpus wallichianus** C. Presl, Bot. Bemerk 110, 1844. Pilger in Engler Pflanzenr. 18 Heft 4(5). 59, 1903. Raizada & Sahni 105, 1960; Dallimore and Jackson. rev. Harrison. Handb. Conif. and Ginkgo. ed 4. 554, 1966. *Podocarpus latifolia* Wall Pl. As. Rar. 1.26.t.30. 1830. FBI V 649, 1888, Brandis Indian Trees; 695. 1906.

WALLICH'S PODOCARP (Pl. 7, Fig. 2)

Tam *Nirambali;* Khasi-*Soplong* Burm *Thitmin ma.*

Description : An evergreen tree up to *c* 24 m in height. *Bark* smooth, mottled brown and white, wood aromatic. *Leaves* opposite or sub-opposite, coriaceous, ovate lanceolate, many nerved, 7.6-17.7 cm long, 1.27-5 cm wide, petiole very short 2 mm long. *Male strobili* in spikes 1.27 cm long, 2-5 together, sessile or on a short peduncle, bract ovate, acuminate, denticulate. *Seed* solitary ovoid 1.9 cm across blue-black seated on a fleshy receptacle.

Distribution : This is the only naturally occuring conifer in S India, Western Ghats from the Nilgiris south wards. Assam, Martaban, Tenasserim usually at altitudes of 900-1500 m. Collected from the interior of Gt. Nicobar Island by K.C. Sahni (1953), its absence from the neighbouring, Andaman Islands is interesting.

Specimens examined : *K.C. Sahni* 22992, hillside above Alexandra River, Gt. Nicobar.

LARGE LEAF PODOCARP (Pl. 7, Fig. 9)

Merrill (1941) records **P. macrophyllus** (Thunb.) D. Don in Kingdon-Ward's collection from North Burma. Shrub or a small tree 7.6-15.2 m tall. Branches strong, horizontal, branchlets dense. *Leaves* densely and spirally arranged, 12.7 cm -17 cm long 1.27 cm wide, thick, leathery, apex pointed or blunt and rounded. *Male* strobili sessile, 3-13.8 cm long. *Seed* ovoid 0.84 cm, scaled on a purple, fleshy receptacle 1.27 cm or more in length.

Amongst the exotics *P. gracilior* Pilger from Ethiopia, from an altitude of 8000′ has done well at Dehra Dun. There is a stately tree at State Forest College. Town Section, Subhash Road, Dehra Dun. 60′ tall with a rounded crown *c* 75 years old and there are several at New Forest, FRI, Dehra Dun *c* 60 years old.

Family 6. *ARAUCARIACEAE Henk. & Hochst. (MONKEY-PUZZLE FAMILY)

Evergreen resinous trees of the southern hemisphere excluding Africa. *Branches* whorled. *Leaves* leathery spirally arranged, subulate or scale-like, stiff (*Araucaria*) to broad ovate (*Agathis*); trees *dioecious* less commonly monoecious (*Agathis*); male strobili large and cone-like, axillary or terminal on short branches with numerous stamens and many celled anthers; pistillate flowers in terminal heads becoming a large woody deciduous cone.

Two genera *Araucaria* and *Agathis* are in cultivation in the subcontinent. This family extends back to the Jurassic period 195 million years (Bayard Hora 1981, p 63).

Leaves awl-shaped or lanceolate; usually dioecious. Male strobili solitary or in clusters. Seed adnate to the ovuliferous scale, wingless or with more or less equal wings .... 1. *Araucaria*

Leaves usually broad and flat; usually monoecious. Male strobili solitary Cone-subglobose to ovoid. Seed free, usually unequally winged .... 2. *Agathis*

1. **Araucaria** Jussieu

Etymology: After a S. American tribe Araucaria, who live in the regions where it grows wild.

Evergreen trees of the southern Hemisphere, occuring in S. America, Australia, New Guinea, New Caledonia, the New Hebrides, and Norfolk Island. *Branches* whorled. Leaves spirally arranged, flat, subulate or scale-like and stiff. Strobili usually dioecious. Male catkins cylindrical, solitary or in clusters, consisting of numerous spirally arranged stamens. *Cones* ripening in two or three years, globular or ovoid with woody overlapping scales, each with a solitary wingless seed adnate to the scale. Three species *A. columnaris* (Forster) Hooker, *A. bidwillii* Hooker and *A. cunninghamii* Sweet are commonly seen introduced in parks and gardens in the sub-continent. The genus has a total of 18 species.

Key to Species
(Cultivated in India)

Leaves broad, flattish, homomorphic.

Leaves spirally arranged, but twisted on their bases so as to throw them into two lateral spreading ranks, almost flat. Habit dome shaped, bark not characteristic .... 1. *A. bidwillii*

Leaves for the most part linear, acicular and awl-like, polymorphic. Leaves of the lower branchlets linear, densely imbricate. Branchlets frequently overlapping each other. Habit not dome shaped; bark not characteristic. .... 2. *A. columnaris*

Leaves primordial or abnormal mostly found on the main stem or on adventitious shoots in the shade of the whole tree foliage, slightly oblique and sharp pointed. Habit not dome shaped; bark in young trees papery, flaking off horizontally, resembling cherry wood bark .... 3. *A. cunninghamii*

1. **A. bidwillii** Hooker **BUNYA BUNYA ARAUCARIA BIDWILL'S ARAUCARIA** (Pl. 8, Fig. 1)

Etymology : The specific name commemorates the British naturalist J.C. Bidwill (1815-1853).

A tree to 45 m with a trunk 0.8-1.2 m in diameter. Crown dome shaped. *Bark* thick resinous. *Leaves* spirally arranged, lance shaped on sterile shoots upto 5 cm long × 0.2-0.5 cm wide, dark green, stiff, apex narrowing into a long stiff point. *Male female* strobili usually on different trees, occasionally on the same. Male catkins up to 15 cm - 17.8 cm long, 0.7 cm wide produced near the points of the branches on the upper parts of the tree. *Cones* erect, borne on the higher branches elliptical or globose upto 30.5 cm × 23 cm wide; scales numerous, 10 cm × 7.5 cm, apex drawn out into a long recurved point. *Seed* large, pear-shaped, 5-6.3 cm long over 2.5 cm wide in the widest part, maturing the third year; wings rudimentary. Native of Queensland, Australia. Ripe seeds are produced each year, a full crop is obtained every third year.

Fl. : Feb.-May (Ootacamund).

Propagation : By seed : by cuttings from young branches; by basal suckers (Macdonald Encyclop Trees 1982) widely introduced in the subcontinent since long.

2. **A. columnaris** (Forster) Hooker *A. cookii* R. Br ex Lindley, *Cupressus columnaris* Forster. **COLUMNAR ARAUCARIA COOK'S ARAUCARIA** (Pl. 8, Fig. 2)

A tree to 61 m. *Branches* horizontal, lateral branchlets long, slender whip-like outer bark shredding off in thin papery layers. Crown shape columnar. *Leaves* on juvenile shoots triangular or lance-shaped, up to 0.7 cm long, pointed; on older shoots broadly ovate, 0.2 cm -0.5 cm long, 0.15 cm wide. Male catkins 2.5 cm -6.3 cm long × 0.7-0.95 cm wide. *Cone* elliptical up to 15 cm long × 11.5 cm wide. Native to New Caledonia and Polynesia discovered by Capt Cooke during his voyage in the Pacific.

Fl. : Feb. - May (Ootacamund).

Propagation: By seed. Widely introduced in the subcontinent. Popular landscape tree. Does will in the Himalaya up to 1200 m.

3. **A. cunninghamii** Sweet **CUNNINGHAM ARAUCARIA** (Pl. 8, Fig. 3)

A pyramidal shaped tree up to 45 m or occasionally 61 m high, with a girth up to 3.6 m. *Branches* long, with branchlets concentrated in dense tufts near the point. *Bark* characteristic, peeling off in thin layers, resembling *Prunus* bark. *Leaves* of two kinds, those on young trees and lateral branches spirally arranged, usually lance-shaped or triangular, 0.42-0.9 cm long sharp pointed; on old trees and coning branches, shorter, crowded and sharp pointed. Male *Catkins* 5-7.5 cm long. *Cones* ovoid, 10 cm × 7.5 cm; scales broadly cuneate with a long awn-like, reflexed apex. *Seed* with a narrow membranous wing on each side.

Fl. : January - March (Ootacamund).

Propagation : By seed. Fast growing, widely cultivated in the subcontinent particularly in Dehra Dun. It does well up to 1000 m in the W Himalaya and up to 1500 m in th E Himalaya.

2. **Agathis** Salisbury nom cons. **KAURI PINES**
Dammara Lamarck

Tall, evergreen, resinous trees with massive columnar trunks confined to Australia, New Zealand, Fiji, Philippines, Sumatra and the Malay Peninsula. Introduced in Dehra Dun, Ootacamund and the highlands of Sri Lanka. *Bark* thick, scaly, resinous. *Branches* horizontal. *Leaves* spirally arranged on the main axis, opposite or alternate on lateral shoots; reddish when young, dark green later, leathery, mostly broad and flat with parallel lines, narrowed into a short, flat stalk. *Male* and *female strobili* usually on the same tree. Male strobili, cylindrical solitary in the leaf axils. *Cones* globose or ovoid, compact, cylindrical; scales fan-shaped, falling as soon as the seeds are ripe. *Seed* solitary, with a well-developed wing on one side. The wood is of good quality. The tree contains resin which exudes from wounds on trunk and branches. One species *A. robusta* has been introduced in large numbers in Dehra Dun where is has done well producing fertile seed. Total species 20.

1. **Agathis robusta** (C. Moore) F.M. Bailey **S. QUEENSLAND KAURI**

(Pl. 9, Fig. 1-4)

Etymology : From the Greek agathis, a ball of twine, since the overlapping scales on the cone look like rolled-up threads; *robusta* means robust.

A large stately tree, native to Queensland, 40-50 m tall with a girth of *c* 4 m, with a straight columnar trunk, free from branches for the greater part of its height. *Bark* thick brown, scaly. *Leaves* leathery ovate or elliptical opposite or subopposite 5-12 cm long, 2-4 cm wide glossy dark green above, paler beneath, with parallel veins, margins thickened, tapering into a broad, flattened petiole. *Male strobili,* cylindrical, 5-10 cm long, c 1.2 cm wide ♀ *cones* sub-globose or ovoid, 10 cm-12.5 cm long, 7.5-10 cm in diameter. Scales closely overlapping 2.5-3.7 cm wide. *Seed* flattened, oblong-ovoid, about 1.25 cm long, with a wing about 2.5 cm long, on one side.

Native to Queensland from sea level to nearly 800 m.

Fl. : Mar-Apr. Male. Arboretum 6, New Forest, *Vaid* s.n. Flowered for the first time 11.4.1955 in New Forest, Dehra Dun. *Fr.* : July. Bunglow 9, New Forest, *K.P. Sagreiya* s.n. ♀ cone formed for the first time 20.7.1961. Some seeds germinate on the cone itself.

Propagation : The seeds are sown soon after the cones mature in July. Germination in Dehra Dun is over 90%. In Dehra Dun it fruited after about 27 years of its introduction from seed. It has been introduced in Dehra Dun and in the Forest Research Institute, Dehra Dun, There are some majestic and stately trees in the latter. It is a good timber for general joinery and is a fine avenue tree.

Family 7. **PINACEAE** Lindl nom. cons.
Lindl. Nat. Syst. Bot. ed 2. 313. 1836.
PINE FAMILY

A large family of resinous trees often gigantic with mostly whorled branches. *Leaves* evergreen (deciduous in *Larix*) stiff, circular, subulate, needle-like, distichous, arranged in 1-7 foliate fascicles, girt at the base by a scarious sheath. *Male* and *female* cones on the same tree (monoecious). Stamens and ovuliferous scales spirally arranged round a common axis, forming terminal or lateral catkins and cones respectively. Stamens numerous, each with 2 anthers, naked, compact, filaments very short; pollen with two bladders (wings) or wingless. *Female strobilus:* ovuliferous scales much enlarged usually numerous, accrescent. *Ovules* two collateral, bract scales also enlarged, narrow, thinner than ovuliferous scales or reduced or hidden. *Fruit* a woody cone, closed until ripe, persistent or shedding the woody scales. *Seeds* two usually unilaterally winged. Cotyledons 3-18. *Seed* with endosperm.

The six genera found in are area are *Pinus, Tsuga, Picea, Abies, Cedrus* and *Larix.* The total world genera are 10 with 250 species distributed in the N Hemisphere southwards to Sumatra, Java, C America and the W Indies.

Key to Genera

Leaves evergreen, pollen grains winged (except *Tsuga*)

  Cones pendulous, woody, persistent

    Leaves acicular, in bundles .... 1. *Pinus*

    Leaves not acicular and not in bundles

      Cone globose not more than 2.5 cm long .... 2. *Tsuga*

      Cone 6-15 cm long, cigar shaped .... 3. *Picea*

  Cones erect, scales sheding off upon the tree, leaving a central axis :

    Leaves distichous .... 4. *Abies*

    Leaves in tufts .... 5. *Cedrus*

Leaves deciduous, pollen grains wingless .... 6. *Larix*

1. **Pinus** Linn. Sp. Pl. 1000 1753 et Gen Pl. ed. 5. 434. 1754. **PINES**

A large genus of 91 species of evergreen resinous trees of which seven species are found in our area. Recently a new species *P. bhutanica* was discovered by the author in 1977 from Arunachal Pradesh and later in 1979 from Bhutan by British botanists. Pines are widely distributed in the northern hemisphere, their northern limit is the edge of the Arctic Circle forming the northern coniferous forests in Europe, Asia and N America, to the subtropical regions of N Africa, the Canary Islands, Asia Minor, Burma, the Philippines, Central America, Florida, the Bahamas, and Hondurus. In tropical countries they are usually found at sub-tropical or warm temperate altitudes, rarely on the plains. *P. merkusii* Jungh et De Vr. of Burma, Sumatra, etc is the only pine in the world that crosses the equator.

The leaves are needle-shaped borne in clusters of two to five. They persist for five years or more. The male and female *cones* are separate but on the same tree. The male cones are short and cylindric, and are catkin like. The female cones have a central axis bearing spirally arranged bract scales, in the axils of which eventually arise the cone scales (ovuliferous or fertile scales) each bearing ovules on the lower surface. The apex of these cone scales becomes enlarged and is called the apophysis. Part of the apophysis may be raised and differentiated to form what is called the umbo. The cones may be erect or pendulous.

Pollination is by wind and pollen is often so abundantly produced that sulphur dust-like pollen grains or pollen showers are visible in the air or on the forest floor. Dispersal of pollen is aided by each grain having two bladder-like wings. They alight on the female cones. At the time of pollination the scales of the female cone open to receive the pollen (sperm). At least a year elapses between pollination in the spring and fertilization during which time the female cone remains small. After fertilization it grows to adult size, at first green becoming brown. The cones mature in about 25 months and open about May in the dry weather to disgorge the winged seeds which take 2-3 weeks to escape completely. Seeds fallen on the ground germinate when the monsoon breaks.

Pines are generally long lived trees. Some specimens of Bristle-cone Pine or Foxtail Pine, *Pinus aristata* of N America are considered to be the oldest living things on this planet, having germinated even before the Indus civilization, some 6000 years ago.

Pinus contains some of the most important timber trees in the world. Chir and blue pine are well knowon and useful joinery woods. They also produce resin of commercial importance and under favourable conditions the revenue derived from resin is more than derived from timber. The resin of Khasi Pine is in quality the most valuable but the areas under this pine are inadequate. In view of its good rate of growth it is gaining importance for afforestation programmes in subtropical areas of the world. The chief value of the Chilgoza pine (*P. gerardiana*) lies in its edible seeds which are roasted and eaten. Pine timber is divided into two main groups; Hard-Pines of the 2 or 3 needled species and the Soft-Pines of the 5 needled species. In general, the heartwood in hard-pines is darker than the sapwood, while in the soft or white pines the colour of the wood is paler and more uniform. In our area, *P. wallichiana,* soft-pine and *P. roxburghii,* a hard pine, yield timber of commercial importance.

Key to Pinus

Leaves in fascicles of 5

Leaves mostly 11-18 cm long, slightly bent, seed 6-10 mm, wing 1.5-3 cm .... 1. *P. wallichiana*

Leaves mostly 8-18 cm, sharply bent, seed 9-19 mm, wingless or rudimentary 0.2 mm. .... 2. *P. armandii*

Leaves mostly 12-28 cm, pendulous from base, seed 6-8 mm, wing *c* 2 cm .... 3. *P. bhutanica*

Leaves in fascicles of three

Leaves stiff less than 15.5 cm long .... 4. *P. gerardiana*

Leaves soft more than 15.5 cm long.

Wings 4 times as long as the seed; umbo of scales with a prominent beak. .... 5. *P. kesiya*

Wings about the length of seed, umbo of scales with a prominent hooked beak. .... 6. *P. roxburghii*

Leaves two in a fascicle. .... 7. *P. merkusii*

1. **Pinus wallichiana** A.B. Jackson in Kew Bull. 85, 1938. Raizada and Sahni 113, 1960. Dallimore & Jackson. Handb. Conif. Ginkgo rev Harrison ed 4, 05, 1966. *Pinus excelsa* Wall ex D. Don *in* Lamb. Descr. Gen **BHUTAN PINE BLUE PINE** (Pl. 10, Fig. 2)

Pinus 2: 5.t. 1824, non Lam. 1778 (nom illeg) Hook.f. FBI 5, 651, 1885. Brandis Indian Trees 689, 1906. *P. griffithii* McClelland in Griff. Notul. 4. 17. 1854 (nom illeg). *Pinus chylla* Lodd. Cat. 50. 1836, nom nud. *Pinus nepalensis* De Chambray, Traite' Prat. Arb. Resin. Conif 342, 1845, non Forbes 1839.

Kail Hindi; *Kail, Yiro* Kashm; Lin, Himachal; *Raisalla, Lamshing* Kumaon; *Neet-kung* Lepcha; *Tongschi, Lamshing* Bhutan, TRADE - *Blue pine, Kail.*

Description : A large evergreen tree, up to 50 m tall. *Bark* greyish brown with shallow fissures. *Leaves* needle-like in bundles of five, 10-20 cm long, drooping, with a bluish green tinge for which it is known as blue pine. Dorsal needle resin canals, 2, marginal. Ventral, 1, symmetric. The male and female *strobili* are borne on the same tree. The male *strobili* (catkins) are 1-2 cm long, ovoid or oblong are borne on the current years young shoots forming a cluster about 5 cm long. The catkin is enveloped by 8-12 overlapping scales. The catkins fall after ripening. The young female cones are erect initially. On ripening they become pendulous, 2-3 together banana shaped, 15-30 cm long, light brown with thin scales adhering the cone, like the scales of a fish and giving a smooth appearence. *Seed* bluish, egg-shaped compressed, 6.5-9 mm × 3.5-5 mm with a wing 1.5-3 cm long, obliquely truncate.

Distribution : From Afghanistan to all along the Himalaya at 1800-3900 m from Pakistan to Arunachal Pradesh through Nepal and Bhutan, very rare in Sikkim and a considerable portion of Kumaon. At higher elevations it is associated with birch and juniper as seen around Gaumukh above Gangotri.

Specimens examined : ARUNACHAL PRADESH, *K.C. Sahni & Naithani* 798 near Bomdilla, *K.C. Sahni* 5030, Ziro; *R.N. Loganey* 4, Thengbo, ♀ cone 26 cm long. BHUTAN, *R. Bedi* s.n.

Phenology : New leaves March-April, attain full size by Aug.-September and shed the next year from May to July.

Male Strobili ripen and shed their pollen from April to June and fall on the ground. The *female strobili* are visible in April, pollinated from April to June with scales open. After pollination the scales close. The cones ripen about September to November of the second year. The winged seeds are carried far and wide when the cones open during the dry weather in October-November. The empty cones remain on the tree for several months.

Birds : The Himalayan Nut-cracker a brown bird spotted with white is a dweller in blue pine forests along with White Browed Blue Fly catcher.

Uses : Railway sleepers, house construction and light furniture. A sweet liquid known as honey-dew is secreted by Aphids infesting the leaves and is eaten.

It stands atmospheric pollution better than most pines. (Dallimore & Jackson op cit. p 507).

Nomenclature : It was for many years known as *P. excelsa* Wall. ex D. Don. According to A.B. Jackson (1938 loc. cit) the rule adopted at the International Botanical Congress Cambridge 1930, rejecting all later homonyms necessitated a change of name owing to the existence of the earlier *P. excelsa* Lam (*Picea abies* (L) Karst. The name *P. nepalensis* De Chamb., was taken up, but it was overlooked that this is itself a later homonym of *P. nepalensis* Forbes (*Pinetum woburnense,* 34: 1839 a name of doubtful application. *P. griffithii* McClelland appeared only as a provisional name, while the analyses on the plate cannot validate the name as McClelland himself says they belong to a different species. The epithet Griffithii could not be applied as Parlatore (in Dc. Prodr. 16, Pt. 2, 411: 1868) had validly published, the combination, *P. griffithii* based on *Larix griffithii* Hk.f.· & Th. (= *L. Griffithiana* Hort. ex Carr.) A new name for *P. excelsa* Wall ex D. Don was therefore required and *P. wallichiana* was proposed by A.B. Jackson.

Pinus wallichiana A.B. Jackson var. **parva** K.C. Sahni **var. nova** fig. 1-5. *Ind. J. For.* 12(1): 40-42, 1989.

Similar to *Pinus wallichiana* A.B. Jacks., but differs in shorter needles 6-11-(13) cm long, smaller ♀ cone 10 cm long, smaller seed 3 mm long and smaller wing 10 mm long.

Type : India, Arunachal Pradesh, Tawang, 10,000′, Kameng District, *R.N. Loganey* 6, 4.1.1959 (holotype DD !).

An evergreen tree. *Leaves* needle-like in bundles of five, 6-11-(13) cm long, margins minutely sinous; apex pointed; needle resin canals, 2 marginal dorsal, one ventral; fascical sheaths of needles *c* 1.4 cm long, very light golden brown. The male and female strobili are borne on the same tree. The male *strobili* (catkins) are *c* 10 mm long × 5 mm broad, ovoid or oblong forming a cluster. *Female cones* terminal, 10 cm long, 3.5 cm broad at the base, gradually narrowing towards the apex, brown, woody cone scales *c* 1.9 cm long and 1.6 cm broad, orbicular ovate. *Seeds* ovoid 3 mm long, wing 1 cm long and 0.5 cm broad.

Distribution : So far, there is a single collection from Tawang at 10,000′ in Kameng District. Therefore it appears to be endemic to Kameng District. Intensive field work is necessary in Kameng and adjacent districts of Arunachal Pradesh, Bhutan, S.E. Tibet and W. China.

Ecology : A temperate rain forest tree, E. Himalayan Wet Temperate of Champion & Seth 1968 (*A Revised Survey of Forest Types of India*). The commonest conifer around Tawang is *Pinus wallichiana* A.B. Jacks., with several rhododendrons, commonly *Rhododendron arboreum* Smith, *R. thomsonii* Hook.f., *R. dalhousiae* Hook.f., *R. falconeri* Hook.f., the last two are uncommon.

Specimen examined : INDIA, ARUNACHAL PRADESH, Kameng District, Tawang 10,000′. *R.N. Loganey* 6, 4.1.1959 (DD !).

Flowering & Fruiting : The male inflorescence starts ripening in January, which is evident from the above specimen. The ♀ cone is ripe with seeds in the 2nd year in January as is evident from *Loganey* 6.

The distinctive features of the var *parva* and its ally *P. wallichiana* are listed in TABLE I, also showing their distribution.

TABLE I

	var *parva*	*P. wallichiana*
Needle length	Mostly, 6-11-(13) cm.	Mostly 11-18-(20) cm
Needle resin canals, dorsal.	2 marginal.	2 marginal.
Needle resin canal	1, median	1, median

canals, ventral.	symmetric	symmetric.
Female cones	10 cm long, 3.5 cm broad	15-30 cm long.
Seed length	3 mm long, wing 10 mm long, 5 mm broad.	6.5-9 mm long, 3.5-5 mm broad, wing 1.5-3 cm long. truncate.
Distribution :	So far, only collected from Tawang, at 10,000′.	From Afghanistan to all along the Himalaya at 1800-3900 m. From Pakistan through Kashmir to Arunachal Pradesh. Nepal and Bhutan. Rare in Sikkim and large parts of Kumaon. China (S. Xizang).

Note : The small size of the female cones is significant. Contrasted with this is the large banana shaped cone of *P. wallichiana.* Reproductive characters in general are much more nearly constant than vegetative characters (Lawrence, G.H.M. p 58, 1951 *Taxonomy of Vascular Plants*). Also Lawrence *loc. cit.* states. 'It has been found from experience in dealing with plants of temperate regions that certain vegetative characters are generally more constant than are others'. The small size of the needles of this new variety fits in with Lawrence's observation as Tawang is located in the temperate belt.

2. **Pinus armandii** Franchet in Nouv. Arch. Mus. Hist. Nat. Paris ser. 2, 7: 95t. 12 (Pl. Davidianae 1: 285, Pl. 12) 1884. Raizada and Sahni 113, 1960. Dallimore and Jackson Handb. Conif. Ginkgo rev. Harrison ed 4, 402, 1966. **ARMAND'S PINE** (Pl. 11, Fig. 2)

A tree distributed in Central and Western China and Taiwan discovered by Pere David in 1873. F. Kingdon-Ward (1952) recorded this pine from Lohit District, Arunachal Pradesh. It has since not been recorded from Arunachal to date where the author carried out several explorations (1969, 1977, 1978). It is apparently absent in Kameng District. Intensive explorations are necessary in the other four districts of Arunachal Pradesh.

Description : A tree 18 m tall and over 0.9 m in diameter, with horizontal branches. *Bark,* thin, smooth pale grey green. Young shoots green covered at first with small translucent glands. *Winter* buds cylindrical, blunt, slightly resinous. *Leaves* in fives, persisting two to three years, spreading or drooping, usually sharply bent 1.3-2 cm from the base, mostly

8-18 cm long, margins finely toothed, apex pointed, needles bright grass green in colour. Inner faces green or whitish. Resin canals 3, 2 marginal and 1 median; basal sheath and scales deciduous. *Cones* subterminal on 2.5 cm long stalks or more, 1-3 together, broadly clindrical, tapering to a rounded apex becoming barrel shaped, 10-17.5 cm long and 5-8.7 cm wide, erect at first, pendulous during the second year; scales thick and rigid woody, resinous, *c* 3.2 × 2.5 cm, broadly triangular, concluded portion, reddish brown, exposed part yellowish brown and rounded at the apex, which is sometimes slightly reflexed. *Seed* 9-19 mm wingless or wing very rudimentary 0.2 mm, liberated soon after ripening. *Male strobili,* smaller, more loosely arranged than in *P. wallichiana.*

This species is closely allied to *P. wallichana,* its wingless seed quickly separates it from the blue pine. The cones of the two species are quite distinct, those of Aramand's Pine being much stouter and scales wider than those of blue pine.

Distribution : Kingdon Ward (1952) recorded it from Lohit District, Arunachal Pradesh. Unfortunately he cites no specimen with specimen number and there is no specimen from Arunachal Pradesh in DD to date. He however writes '*P. armandii,* Blue Chinese Pine with very long needles at 7000′, weeping habit associated with *Tsuga dumosa, Picea spinulosa* (*P. morindoides*), *Rhododendron arboreum, Populus ciliata,'* etc. (p 207, 211). Fletcher (1976, p 276) reports it from Trulung, Gorge of Tsangpo "dense temperate rain forests being on all sides with open woods of *P. armandii"* collection number for this pine is not cited, although most of the specimens mentioned in Fletcher's account bear collector's name and number. The author failed to locate any trees in Kameng District inspite of extensive explorations (Sahni 1964, 1977, 1978). In China it usually occurs on rocky terrain in mountains of W. China, Yunnan Hupeh, Szechwan, south of Yellow River in Shansi. NE Frontier Burma (Lace, p 169 1922).

Specimens examined : CHINA, W. Hupeh, *E.H. Wilson* 2509 (DD) Expedition to China 1907-1909.

The seeds are edible (Kingdon-Ward, J.R. hort. Soc., 1952, *77,* 207).

3. **Pinus bhutanica** Grierson, Long & Page Notes R.B.G. Ed in 38(2): 297-310 (1980). **ARUNACHAL PINE** (Pl. 10, Fig. 3)

This five needle pine was discovered by the author on 6th April 1977 from Tenga Valley, Kameng District, Arunachal Pradesh, and incorporated in the Forest Research Institute, Herbarium, Dehra Dun, *Sahni & Naithani* ser II, 381. Later in 1979 it was also found by Grierson and Long from Bhutan and described by them along with C.N. Page as a new species. Notes R.B.G. Edin. loc. cit.

As it was first collected from Arunachal, I have given it the appropriate name Arunachal Pine, as the blue pine is also popularly called the Bhutan Pine, from where it was first collected and described.

Sahni 381, was sent to Dr. Page for comparison with known Chinese five-needle pines. Grierson et al loc. cit. p. 299 commented. "Independently Sahni of the F.R.I. Dehra Dun, recognised a pine at Tenga in the neighbouring Arunachal Pradesh State of N.E. (formerly NEFA) with long needles but winged seeds and cones like those of *P. wallichiana,* the identity of which he rightly questioned, noting also its differences in needle anatomy from *P. wallichiana*".

Description : Tree *c* 25 m tall. Branches spreading, drooping. *Bark* smooth, thin, pale grey green, wrinkling when dry. *Leaves,* needles 5, mostly 12-28 cm long, outer face gras green, inner faces strongly whitish bloomed. Needle posture pendulous from base. Needles are shed during second year. Needle resin canals, dorsal, 2, marginal or submarginal. Neelde resin canals ventral, one rarely two, marginal or submarginal, asymmetric. *Cone* shape elongate-cylindric 12-20 cm long. Cone peduncle 4.5-6 cm long. Cone scales thinly woody ± flexible, apex flat. *Seed* 6-8 mm, long wing *c* 2 cm.

Distribution : Tenga Valley, Kameng District, Arunachal Pradesh at *c* 1700-2000 m Eastern Bhutan.

This pine has a characteristic crown shape, it has a much narrower crown than the blue pine which has a wide spreading crown. The needles 12-28 cm are much longer than those of either *P. wallichiana* or *P. armandii.* The fascicle sheaths of the needle are also long *c* 1.8-2.3 cm, bright golden brown in colour.

The corresponding log of Sahni 381 was collected by felling the tree for study of its wood anatomy by the Wood Anatomy Branch, Forest Research Institute, Dehra Dun. It is incorporated in the Wood Anatomy Branch, Xylarium collection.

4. **Pinus gerardiana** Wall. ex. Lamb; Genus Pinus II. : :5, t 79, 1832. Hook.f. F.B.I. 5, 652 1882. Brandis Ind. Trees 690, 1906. Raizada & Sahni 114, 1960. Dallimore and Jackson-rev-Harrison Hand. Conif. Ginkgo. ed. 4. 429, 1966. **CHILGOZA PINE** (Pl. 11, Fig. 1)

Chilgoza, Neoza (Seed) Hindi, *Chiri, Gunober Prita* N.W. Himalaya.

Etymology : It was discovered by Capt. Gerarad in the early 19th Century after whom it was named.

Description : An evergreen tree 15-18 m occasionally up to 24 m high, usually 1.8-2.4 m in girth. *Bark* silver grey, thin, smooth, peeling off in large thin flakes. *Leaves* in threes, 5-10 cm long, stiff, dark green. *Male* catkins 7.6-13 mm long. *Female* cones near the middle of the shoot. Mature cones obong-ovoid 15-23 cm × 10-13 cm. Scales 3.8 × 2.5 cm, thick and woody with a recurved beak. *Seed* 2.5 cm long, cylindric oily with a short caducous wing.

Distribution : NW Himalaya with scarce rainfall but heavy winter snow fall. Unlike chir the chilgoza is not found in places with a monsoon rainfall. Local in the inner semi-arid valleys from Bashahr westwards to Kashmir, Chitral, N Baluchistan and Afghanistan between 1800-3,000 m.

Specimen examined : AFGHANISTAN : *J.E. Aitchison* 113 Kurram Valley KASHMIR *Dr Giles s.n.* 1887, Gilgit expedition; HIMACHAL PRADESH, *R.N. Parker* 2870, Bashahr.

Phenology : Leaves evergreen which fall by the second year. New leaves in April-May.

Flowers : May-June, when pollination takes place. The young female cones increase in size rapidly duing the second years, ripening about Sept.-Oct. of the year after. The seeds are not viable for long.

Uses : The seeds are its chief economic product and prevent exploitation of the timber. They are exported to the plains in the winter.

5. **Pinus kesiya** Royle ex Gord. Gardn. Mag. XVI, 8, 1840. Dallimore & Jackson rev. Harrison Handb. Conif. Ginkg. 436, 1966. *P. khasya* Royle Gdn. Mag 1840, 8. Hook.f. F.B.I. 5, 652, 1882. Brandis Ind. Trees 690, **KHASI PINE** (Pl. 10, Fig. 4)

1906. *Pinus insularis* Endl. Conif. 157, 1847. Raizada and Sahni 115, 1960.

Ding-se, Dieng-kysi, Khasi; *Far* Lushai; *Saral* Beng; *Tinyu, Tinshu* Burm.

Description : A small or moderate-sized tree with a rounded crown, a smaller tree in Khasi Hills but attaining 30-45 m height in Burma. *Branches* whorled. *Bark* thick deeply and reticulately cracked, peeling off in thick plates. *Leaves* bundles of three, 15-25.5 cm long slender, apex a long fine point. *Cones* solitary, in pairs or occasionally in three, ovoid, symmetrical, bright brown, 5-7.5 cm × *c* 5 cm across, stalks short and stiff, scales of cone in spirals of 8-5, thickened towards the apex, with a blunt knob at the end. *Seeds c* 8 mm long with a round topped wing, wing 4 times the length of the seed.

Distribution : Khasi and Naga Hills, Manipur, Upper Burma and the Philippines. It occurs at elevations of 800-2000 in Khasi Hills where it thrives at elevations of 1200-1400 m.

Specimens examined : KHASI HILLS, *D.C.F.* s.n. NAGA HILLS, *Jagermani* 779 Imphal.

Phenology : The new leaves appear in Feb. - March. The old needles fall in April-May, by which time the new ones are full sized. The needles persist for about 1 year or 15 monghs. The male and female strobili appear in Feb.-March. The female strobili 1-3 at the apices, are *c* 1-3 × 0.9-1 cm in diameter, light green, ovoid. By next year the cones are nearly full sized but still green; they ripen from Feb.-April, two years after the appearance of the female strobili and persist long on the tree.

Uses : The quality of resin is valuable. It is a fast growing pine and is in demand for afforestation in subtropical countries of the world as well as on hills of S. India.

6. **Pinus roxburghi** Sargent in Silva N America 2, 9/1897. Raizada & Sahni 117, 1960; Dallimore & Jackson. Handb. Conif. Ginkgo : rev. Harrison 432, 1966. *P. longifolia* Roxb. Fl. Ind. III : 651, 1832, H.K.f. FBI 5, 652, 1882. Brandis Ind. Trees 690, 1906.

CHIR PINE, HIMALAYAN LONG NEEDLE PINE

Chir. Chil Hindi; *Dhup* Nep; *Nyit,* Lepcha

Description : A large evergreen tree nearly deciduous in dry localities, usually not exceeding 30 m in height, rarely 55 m and a girth of 3 m have been recorded. *Branches* whorled up to middle age, crown elongated to pyramidal shaped, afterwards becoming spreading or umbrella shaped. *Bark* 2.5-5 cm thick, exfoliates in long plates, brownish-red, turning light grey. *Leaves* in bundles of 3, 20-33 cm long, bright green, each bundle covered at the base by a persistent sheath. *Strobili* Monoecious. Male catkins 1.3 cm long. *Female* cones solitary or 2-5 together. Cones 10.2-20 cm × 7.5-13 cm, with thick woody scales, the scales pyramid-shaped with a pointed or curved beak. *Seed* 7.5 mm- 1 cm long with a long thin membranous wing.

Distribution : All along the Himalaya from Pakistan to Arunachal Pradesh at 450-2300 m in the outer ranges where the full force of the monsoon is felt. Absent in the Kashmir Valley proper where the full force of the monsoon is not felt.

Specimen examined : K.C. Sahni 445, Munna, Kameng District, Arunachal Pradesh; Bletting along Tawang Cho. *R.N. Loganey* 13, 24.12.1956; Rashing to Dirang *R.N. Loganey,* 11 ♀ cone in Fr. Museum, D.D.

Phenology : New needles Feb.-March, persist for about one year and 5 months. Shed May-June. *Flowers* Feb.-April. Seeds are shed in April-May. At low elevations the male inflorescence is visible in early January. The flowers ripen, and pollen showers take place from Feb.-April according to altitude. The reproduction of conifers in general is best examplified by chir pine. The male inflorescence becomes visible in January. As it matures, the scales open, showering in abundance, sulphur dust-like pollen grains. Each pollen grain has 2 balloon-like wings which aid dispersal by wind and alight on the female cones. A year elapses between pollination in the spring and fertilization. At the time of pollination the scales of the female cones open to receive the pollen. The female cones grow fast to adult size after fertilization, the scales thickening, bending back and closing tightly. They mature in 25 months and open about May in the dry weather to disgorge the winged seeds. The fertilized winged seeds are carried for and wide by the wind. Seeds fallen on the ground germinate when the monsoon breaks, some are eaten while on the tree by flying squirrels.

Uses : It produces resin of commercial importance. It is a useful joinery timber and also used as **railway sleepers** after treatment.

7. **Pinus merkusii** Jungh et. De Vr. in PL. Nov. Ind. Bat. Orients, t.2 1845. Hooker.f. FBI 5, 652 1882. Raizada & Sahni 118 1960. Dallimore & Jackson. Handb. Conif. Ginkgo. rev. Harrison 443 1966. Brandis Indian Trees 691, 1906. *Tinyu, Tinshu, Shaja* Burm. — **TWO NEEDLE PINE, MERKUS PINE** (Pl. 11, Fig. 3)

Description : A moderate sized to large tree attaining 18 m in Burma and 30 m in Sumatra, of conical habit when young; spreading or round headed when mature. *Leaves* in bundles of two, 15-25 cm. *Cones* usually in pairs 5-7 cm long, cylindric-conical, erect, peduncle 1.3 cm long, scales with a thick flat pyramidal beak. *Seeds* small many times shorter than the unequal sided wing.

Distribution : In Burma it occurs on low hills and spurs at elevations of 150-760 m associated with *Dipterocarpus tuberculatus.* It extends from the southern Shan States southwards through the hills of the Salween and Thaungin drainage, also found in Thailand, Cochin China, Sumatra, Java, Borneo and the Philippines. It is the only Pine in the world that crosses the equator. Recently collected from Lohit Dist., Arunachal Pradesh.

Specimens examined : BURMA, *J.H. Lace* 4720, Thaungin Valley; *C.G. Rogers* 332, Maungpok peak 3500′.

Phenology : New *leaves* Feb-March. *Cones* ripen April-May.

Uses : General purpose construction timber. It produces resin of good quality.

2. **Tsuga** Carr. — **HEMLOCK**
Carr. Traite Conif. 185. 1855.

The Hemlocks as they are popularly called are native of N. America, China, Japan, Taiwan and the Himalaya with 10 species of which 2 are in our area. Evergreen tree with massive trunk with horizontal branches. Branchlets often pendulous at the tips. *Leaves* flattened linear, usually two ranked with 2 whitish bands beneath. Leaf-stalks slender, short, adpressed to the shoot seated on a tuberculate cushion. There is a solitary resin canal below the midrib. *Male and Female strobili* on the same tree. Male catkins globose, axillary. Female cones small terminal on lateral shoots, ovoid with persistant woody scales, ripening in one season. There are two species in the area covered in this work. Small female cones up to *c* 2.5 cm long is the outstanding diagnostic feature of this genus of the *Pinaceae.*

Key to Tsuga

Margin of leaves minutely toothed. Young shoots light brown.

Leaves 2.5-3.1 cm long, tapering to the apex, dark green above with two bands of silvery-white stomata below. Cones ovoid up to 2.5 cm long .... 1. *T. dumosa*

Margin of leaves entire; young shoots reddish grey, hairy.

Leaves 6-19 mm long, rounded or sharp-pointed, stomatic bands on the undersurface conspicious. Cones ovoid 1.2-1.8 cm long .... 2. *T. yunnanensis*

1. Tsuga dumosa (D.Don) Eichler in Engler u. Prantl. Pflanzenfam 2(1): 80. 1887. Raizada and Sahni 119, 1960; Dallimore and Jackson Hanbd. Conif. & Ginkgo. rev. Harrison ed 4. 641, 1966. *Pinus dumosa* D. Don Prod. Fl. Nepal 55, 1825. *Pinus brunoniana* Wall. Pl. As. Rar. 3. 247. 1832. *Tsuga brunoniana* (Wall) Carr. Traite Conif. 188. 1855. Hook.f. F.B.I. : 654. 1888. Brandis Indian Trees 693, 1906.

HIMALAYAN HEMLOCK

INDIAN HEMLOCK

FIR

(Pl. 13, Fig. 2)

Changathasi, Thingia, Sula Nepal; *Tangshing* Bhutia.

Etymology : From the Japanese name for the Asian species.

Description : A tall tree up to 36 m high with a massive trunk up to 8.5 m. in girth have been recorded. *Bark* thick, rough. Branches drooping. *Leaves* 1.27-2.54 cm, more or less 2 ranked, needle-like, silvery white beneath. *Male & Female strobili* on the same tree. Male catkins 0.25-0.5 cm long. Stamens with 2 rounded pollen sacs. Ripe female *cones* 1.3-2.5 cm long, ovoid, scales rounded shining, leathery, persistant. *Seeds* 0.78-0.88 cm long, wing oblong, obtuse.

Hemlock is like Fir in foliage, but is quickly separated by much smaller female cones (up to 2.5 cm) in the former. For its resemblance with fir it is also called Hemlock Fir. Another characteristic field character is its massive trunk of wide girth.

Distribution : E. Kumaon through Nepal, Sikkim, Bhutan to Arunachal Pradesh and Upper Burma, at 2400-3000 m. Tibet (Jingwei, 1982 p 20).

Specimens examined : KUMAON, *R.N. Parker* 2055, Tithla, Almora Dist., NEPAL, *N.D. Bachkheti* s.n. Arun Valley, E. Nepal; SIKKIM, *K.C. Sahni* 32, Lachen, 120′, 15′ g.b.h.; *Jagit Ram* s.n. Lachung. ARUNACHAL PRADESH, *K.C. Sahni* 5181, below Peri La, Kameng District, *K.C. Sahni* 461, Se La; BURMA, *C.W.D. Kermode* 17160, Myitykina 9000'.

Phenology : Fl. May - June. *Cones* ripen the same year.

Propagation : By seed in the spring or from cuttings of mature wood by end of June.

Uses : Timber made into shingles, thick rough bark used for roofing.

Mehra & Jain (1976) have contributed on its leaf anatomy and histology of the scale and female cone.

2. **T. yunnanensis** (Franchet) Masters. Raizada and Sahni 119, 1960. Dallimore & Jackson. Handb. Conif. rev. Harrison 648, 1966. **YUNNAN HEMLOCK** (Pl. 14, Fig. 1)

Abies yunnanensis Franch. in Journ de Bot. 13: 258, 1899. Chinese *Tich Sha.*

Description : A tree 30 m or more in China with a massive trunk up to 6 m girth. Young shoots reddish grey, covered with dense stiff hairs above. *Leaves* 1.27-1.9 cm long, with 2 prominent white bands of stomata beneath. *Cones* ovoid 1.3-1.9 cm long; scales few, roundish, slightly reflexed at the apex.

Distribution : Western Yunnan and W Szechuen from 2130-3048 m. Tibet (Jingwei 1982, p 20) Forest recorded it from N E Upper Burma. (Orr, M.Y. Pl. Chin. Forrestiawae, Notes R.B.G. Edin 18; 134, 1933.

3. **Picea** Dietrich. Diet. Fl. Gen. Berl. 2: 794. 1824. **SPRUCES**

'Spruces are widely distributed in the temperate regions of the N Hemisphere found in Europe, Asia minor, the Caucasus, Siberia, China,

Japan, the Himalaya and N America'. *c* 60 species have been described perhaps two third have been or should be reduced to synonmy according to Dallimore & Jackson (1966, p 328).

Tall evergreen trees with needle-like leaves with sharp points, arranged singly in spiral order. *Leaves* tetragonal or flattened. They are characterised by the presence of peg-like leaf bases on the twigs. *Male and Female Strobili* on the same tree, appearing in spring in leaf axils of previous year's shoots. Male Strobili, catkin like usually solitary, consisting of numerous overlapping antheriferous scales. Each scale with two linear pollen sacs attached to the margin; pollen grains with two inflated wings. Female strobili solitary at tips of the twigs of the previous year. *Ovuliferous* scales numerous, wide, much larger than supporting scales; ovules two at the base of each scale. *Cones* pendulous, ripening during the first autumn when they open and liberate the seeds, but remaining on the tree for some time after the seeds have fallen. *Seeds* small with a well developed wing. Two species are well known from our area, *Picea smithiana* (Wall.) Boiss., the W Himalayan Spruce and *P. spinulosa* (Griff.) Henry, the E Himalayan Spruce. A third *P. brachytyla* was recorded from Kameng District, Arunachal Pradesh (Sahni 1962) some 2 decades back, also from Bhutan and a fourth *P. farreri* from Upper Burma was described recently by Page and Rushforth (1980).

Key to Picea

Needles 2.5-3.8 cm, dark green quadrangular in section, with 2 lines of stomata on all 4 sides. Cone 10-17.5 cm × 2.5-5 cm. . . . . 1. *P. smithiana*

Needles 1.9-3.2 cm, green on dorsal surface, flat in cross section, with 2 lines of stomata on the ventral surface. Cones 10 × 2.5-3 cm. . . . . 2. *P. spinulosa*

Needles 1.27-1.9 cm, green to yellowish green, flat in cross section, conspicuously bluish-white beneath with 2 broad bands of stomata. Cones 6.3-8.8 cm. . . . . 3. *P. brachytyla*

Needles (1.5-) 1.8-2.3-(2.5 cm), blue green, flat in cross section bright niveous white below, with 2 bands of 5-6 rows of stomata below. Cone (6-) 7-9.5 (-10) × 3-4 cm. . . . . 4. *P. farreri*

1. **Picea smithiana** (Wall.) Boiss. Fl. Orient. V: 700, 1884; Dallimore & Jackson, Handb. Conif. ed. 3. 347.f.81. 1948; Raizada and Sahni 121 1960. *Pinus smithiana* Wall. cat. 6063, 1827 et Pl. Asiat. Rar III:

WEST HIMALAYAN SPRUCE
(Pl. 15, Fig. 3)

24 t, 246, 1832. *Picea morinda* Link in Linnaea XV, 522, 1841. Hook.f. FBI 5: 653 1888. Brandis Ind. Trees 692, 1906.
Rai, Rau, Riar, Kachlu N.W. Him.; *Roi, Ragha, Morinda* Garhwal & Kumaon.

A very large evergreen tree of pendulous habit up to 60 m and above and a girth of as much a 5.7 m. *Branches* whorled, horizontal, branchlets hanging like tassels. *Bark* rough, cut by shallow furrows into small plates. *Leaves* arising singly and spirally around the branches, 2.5-3.8 cm long, dark green, needle like, sharply pointed, 4 sided, with *c* 2 lines of stomata on all 4 sides, stiff. *Male & Female* strobili on the same trees. Male catkins 2.54 cm long, solitary, erect. Female strobili in *cones* which are solitary, terminal cylindric, tapering at each end 10-17.5 cm x 2.5-5 cm pendulous dark brown. *Seed* 0.5-0.6 cm. dark-grey or blackish, wing spoon-shaped, total length of seed with wing 1.27 cm x 0.7 cm; wing light brown.

Distribution : In the Himalaya from Afghanistan to Kumaon at 2,150-3,300 m associated with fir, deodar, blue pine and at higher elevations with *Quercus semecarpifolia,* Maple, etc.

Phenology : New needle appear in April. They persist for 3 to 6 years, shed mostly in May-June. Pollen is shed from mid-April to mid May. Soon after pollination the scales close and fertilization is completed. By July the cones reach full size though still green ripening in the second year.

Uses : Planking in ceilings and floors. Bulk of the supply goes to the railways as treated sleepers. Suitable for glider construction and was in use by air-craft establishments in India. Suitable for production for newsprint.

A huge tree of immense height and girth is seen near Mundali, Chakrata Division, Garhwal. It is protected by the Forest Dept. unfortunately the author did not record its height and girth.

2. **Picea spinulosa** (Griff.) Henry in Gard. Chron. ser. 3. 39. 219. 1906. Raizada and Sahni 122 1960.

EAST HIMALAYAN SPRUCE
(Pl. 15, Fig. 1)

Dallimore & Jackson rev. Harrison Handb. Conif. Ginkgo edn 4. 382. 1966. *Abies spinulosa* Griff.

Journ. Travels 259. 1847. et Icon. Pl. Asiat 4 t. 363, 1854. *Picea morindoides* Rehder in Sargent, Trees and Shrubs 1: 95. t. 48. 1903 Vern. *ehsing.*

Description : A tall evergreen tree, one of the tallest of the Spruces attaining a height of over 60 m/200′ in the Eastern Himalaya. *Bark* rough scaling off in small squarish plates. *Branches* more pendulous (than in *P. smithiana*). Young branchlets slender, glabrous yellow grey. *Buds* ovoid, 1.2 cm long, pale brown slightly, resinous. *Leaves* more or less irregularly arranged round the shoot, crowded, closely overlapping on the upper side, flattened, 1.9-3.2 cm long, 1 mm broad, keeled on both surfaces, tipped with a sharp point, dorsal surface green, ventral with two lines of stomata. *Male strobilus* pink 1.9 cm long. *Cones* cylindric, blunt, 5-10 cm × 2.5-3 cm, green with purple margin to the scales when young, shining brown when ripe, scales 1.2-1.5 cm wide. *Seed* including the wing 1.27 cm long, greyish-brown; wing broadest above the middle rounded or denticulate at the apex.

This spruce was discovered by Griffith about 1841 in Bhutan at 8500-10,000′. In the Lachen Valley in Sikkim, one fallen specimen measured 220′ from the roots where it had broken off (Dallimore & Jackson op cit p. 382).

Distribution : SIKKIM, Chumbi Valley (Tibet), Bhutan to Arunachal Pradesh.

Specimens examined/collected : N. SIKKIM, Lachen *K.C. Sahni* 33, 27.7.1960. Tree 70′, 9000′,; Lachen, *Ribu & Rhoomo* 5511. 8000′, 6, Sept., 1911.

3. **Picea brachytyla** (Franchet) Pritzel in Bot. Jahrb. 29: 216, 1900. K.C. Sahni *Indian Forester* 88(10): 748-749, 1962, one plate. Dallimore & Jackson. Handb. Conif. Ginkgo. rev. Harrison edn 4, 346, 1966. *Abies brachytyla* Franchet in Journ. de Bot. 13: 258, 1899. (Pl. 15, Fig. 2)

A tree 10.6 m-30 m high, conical in habit when growing in woods, but commonly round-headed in more open country. *Bark* pale-grey brown. *Branches* relatively long and horizontally spreading, ascending at the tips.

Branchlets pendulous. *Young* shoots slender, pale-brown or buff-coloured, shining, glabrous, or occasionally hairy, becoming darker with age. *Buds* numerous, small, ovoid, *c* 0.4 cm long, with rounded, chestnut-brown closely-pressed scales. *Leaves* crowded and closely overlapping on the upper side of the shoots, those on the lower side spreading outwards in two opposite ranks and exposing the shoots, often rigid, green to yellowish green, 1.27-1.9 cm long, flat, but slightly ridged on each surface, conspicuously bluish-white beneath with two broad bands of stomata; apex bluntish or with a sharp, horny point. *Cones* cylindric oblong, 6.3-8.8 cm long, greenish, tinged with purple when growing, dull broan when mature. *Scales* broad and rounded at the summit or more or less narrowed to a blunt, wavy apex; *seeds* 3-6.4 mm long with a slightly longer wing.

A specimen of this spruce no 22 was collected in 1959 and sent to the author by Sri R.N. Loganey, Assistant Conservator of Forests, from Mago, Kameng District, North East Frontier Agency now Arunachal Pradesh. This was identified by the author (see Sahni loc. cit) as *Picea brachytyla* (Franchet) Pritzel and confirmed as such by Kew, where it was sent for comparison with thier Sino-Himalayan collections.

Distribution : This handsome spruce which may be recognised among the flat-leaved species by its pale brown shoots, numerous chestnut-brown ovoid buds, leaves bluish-white beneath, occurs over an extensive area in Western Hupeh and W. Szechwan from 1500 m-2280 m and in W. Yunnan it covers a wide range of altitude from 2100 m-3700 m. Merrill (1941) identified this species in a collection by Kingdon-Ward from Adung valley in Upper Burma. This find found growing in Mago, Kameng division of North East Frontier Agency is an interesting new record for India and is an evidence of the long recognized affinity between the floras of the Eastern Himalaya and that of Western China, a country which is one vast intricate series of mountain ranges not comparable to the giants of the Himalaya in height but never the less of commanding altitude and even more extensive. In Kameng division it usually occurs in pure patches but is sometimes seen mixed with *Pinus wallichiana* and *Abies densa* as in Mago, *Tsuga dumosa* and *Taxus baccata* as in Thingpo (Sahni 1962). Recently found in Bhutan from Bumthang (Grierson and Long 1983).

This species closely resembles the Sikkim spruce, *Picea spinulosa* (Griff.) Henry and can be separated by the following characters : (Sahni 1962).

Picea brachytyla	*P. spinulosa*
Branches horizontal, branchlets pendulous.	Branches and branchlets pendulous.
Leaves 1.3-2 cm, flat, slightly ridged on dorsal and ventral surface, bluish white beneath.	Leaves 2-2.64 cm, flat keeled on both surfaces, green beneath.
Buds chestnut-brown.	Buds pale brown.
Seed 0.3 cm-0.64 cm, wing slightly longer than seed.	Seeds 0.3 cm, wing 1.3 cm.

Herbarium material of this species is deposited in the Herbarium, Forest Research Institute, Dehra Dun with the following annotation Loc. Mago, Kameng Division. North East Frontier Agency. Alt. 3,500 m, Nov. 23, 1958, *R.N. Loganey* No. 22.

Uses : In Kameng division it is used for constructional purposes. In W. China it has been ruthlessly cut for timber and is now much rare than in former times.

Acknowledgement is made to Dr. G. Taylor, the then Director, Royal Botanic Gardens, Kew, for kindly confirming the determination of this species.

4. **Picea farreri** Page & Rushforth Notes R B G Edin 38(1): 129-136, 1980 (fig. 1, a-f). **FARRER'S SPRUCE** (Pl. 16, Fig. 1)

Etymology : named after R. Farrer who first collected it in the 'Valley of the Winds and Water' (Feng-Shui-Ling Valley).

A tree with broad open crown and highly pendulous ultimate branchlets closely resembling *P. spinulosa* in habit but with shorter more flattened leaves which are niveous white below as in *P. brachytyla* (Page & Rushforth op. cit. p. 130).

Type : Upper Burma Feng-Shui-Ling Valley, 1919, *Farrer* 1435 (holo E, cones and seeds only).

Other specimens : *Kermode* 17216 (DD & K) Loc. Fengshuiling Pass 8500′. A black-and-white photograph taken in the wild and deposited by

Kermode in the Photo Section, Forest Research Institute, Dehra Dun examined by Page is of this species. Page & Ruthforth have given a detailed description of this taxon. Below is a detailed summary of their description.

Description : Tree to 30 m or more. *Crown* broadly conical of open habit. Branches spreading to slightly descending, slender. *All Branchlets steeply descending; the ultimate markedly pendulous. Bark* grayish, scaly. *Shoots* olive-brown to pale orange-brown, dull to scarcely shining. Strong shoots with moderate ridges and grooves. *Foliage* moderately sparse, with *leaves arranged almost all round shoots,* becoming dense above and pectinate beneath on stronger shoots. *Leaves* on all shoots forward pointing, blue green, slightly bloomed, with two bands of 5-6 rows of stomata below, *bright niveous-white* over *almost the entire lower surface* (white colour persists in herbarium specimens). (1.5-) 1.8-2.3 (-2.5) cm long, dorsi-ventrally flattened, slightly ridged both above and below, *parallel-sided, narrow* one mm in breadth for more than 4/5 of their length, tapering for final 2-3 mm to end in acute pungent apex. *Buds* ovoid-conic, chestnut brown, shining, with white resin patches on lower scales. *Female* cones sessile or on less than 0.5 cm stalk, mid-brown, dull to shining, ellipsoid-cylindric, woody (6.0-) 7-9.5 (-10.0) × 3-4 cm when open. Scales 0.8-1.2 × 1-1.6 cm with irregularly rounded to somewhat pointed apex. *Seed* (including wing) *c* 1.6 × 0.5 cm; wing pale brown, papery; seed portion 3.4 mm × 1.5-2 mm, mid-brown to blackish-brown, pointed. *'Distinguishing features* of *P. farreri, P. brachytyla* and *P. spinulosa'* According to Page & Rushforth p. 132. *P. farreri* is separated from *P. brachytyla* by (1) its more pendulous ultimate branchlets (2) more open crown (3) more strongly radial arrangement of leaves around the shoots (4) leaves not closely adpressed above the shoot nor swept down as in *P. brachytyla* (5) leaf less pungent. (6) ♀ cones more delicate with much more rounded, less toothed, less wavy cones scales which become more woody at tip and much smaller. *P. farreri* is separated from *P. spinulosa* by its (1) broader leaves with conspicuous bright niveous-white undersides (2) shoot darker varying from pale orange-brown to olive brown, scarcely shining and pubescent (compared with the very pale, shining and usually glabrous shoot of *P. spinulosa*). (3) sessile pulvinus to each leaf, long, slender in *P. spinulosa* (4) leaf section less quadrangular (5) paired resin canals in most sections, absent from most sections in the latter.

According to Page & Rushforth (op.cit p. 134), it has impressive appearance in the wild. Herbarium annotations indicate Trees over 100′/30.48 m. Kermode (1945) records its altitude as 8000′/2438 m (Farrer) and 8500′ (Kermode) from Feng-Shui-Ling Valley.

Abies Miller **SILVER FIRS**

Mill. Gard. Dict. 1754.

Evergreen trees comprising 40 species confined to the temperate regions of the northern hemisphere, in Europe, N. Africa, Asia from the Himalaya northwards and N. America. Young trees conical or cylindrical. Branches more or less regularly whorled. *Bark* of young trees smooth or beset with resin blisters, furrowed in trees. *Leaves* variously arranged on the branchlets, according to species, those on lateral shoots either pectinate, spreading all round the shoot or crowded; on leading shoots widely spreading. Dark green, pale green or glaucous, linear, rounded or short pointed, often notched at the apex, flattened, marked below with 2 waxy bands of stomata. Bruised leaves emit an odour of turpentine. The needles persist several years and on falling, leave circular leaf scars on the shoots. ***Male and Female Strobili;*** borne on the same tree in spring. Male strobili in short catkins arising from the leaf axils. Stamens spirally compressed, shortly stalked, with 2 globular pollen sacs; pollen grains with 2 inflated vesicles. Female strobili cone-like; ovuliferous scales, spirally arranged, bearing two ovules. *Cones* erect. This genus is distinguished from other conifers by erect cones which break up on the tree when ripe and leaving behind the erect spindle-like axis. *Seeds* with resin vesicle and a permanent wing.

Etymology : From the Latin *abire,* to go away, used here in the sense of distance from the ground to refer to the height some species attain.

There are four species of Abies in the Himalaya :

Abies pindrow Royle (*A. pindrow* Spach), *A. spectabilis* (D. Don) Spach, *A. densa* Griff. ex Parker and *A. delavayi* (Van Tiegham) Franchet.

Key to Abies

Crown narrowly cylindrical, branches short, branchlets glabrous; leaves 2.5-6.2 cm. Female cones 10-18 cm long, perfectly cylindrical; scales fan-shaped, 2.5-3 × 3-3.8 cm, prominently eared. . . . . 1. *A. pindrow*

Crown wider, not cylindrical, ± pyramidal; branches longer, slightly upturned; young branchlets hairy in the grooves, becoming hairless; leaves 1.3-2.5 cm (W. Himalaya), up to 6.2 cm (E. Himalaya). Female cones *c* 7 cm, nearly cylindrical (much larger in cultivated trees) carpellary scales or bracts hidden by the placental or cone scales, and invisible till the cones break up. 2. *A. spectabilis*

Crown wide spreading, umbrella or pagoda shaped with wide horizontally spreading flattened branches; leaves crowded dense, 1.3-5 cm, *margin not revolute.* ♀ cones 10-17.5 cm, cylindric; carpellary scales or bracts with long acuminate tips, which appear between the placental or cone scale . . . . 3. *A. densa*

Crown wide spreading with horizontally spreading *massive* branches and flat tops; young shoots yellowish or reddish brown slightly grooved, or not grooved, slightly hairy; leaves 2-3.1 cm, *margin revolute.* ♀ cones 6.2-10 cm, broadly cylindrical or *barrel shaped;* carpellary scales or *bracts exserted,* often recurved. . . . . 4. *A. delavayi*

1. **Abies pindrow** Royle Ill. t. 86. Fig. a.c. 1836 et p 350, 1839. Raizada & Sahni 127, 1960. *A. pindrow* Spach Hist. Veg XI, p. 423, 1842, Hook.f. FBI: 5, 692, 1890. Brandis Indian Trees 692, 1906; *A. pindrow* (Royle) Spach. Dallimore & Jackson Handb. Conif. Ginkgo. rev. Harrison ed. 4, 74, 1966. *A. webbiana* Lindl. var. *pindrow* Brandis; *Pinus pindrow* Royle.

W. HIMALAYAN FIR
LOW LEVEL FIR
PINDROW FIR
(Pl. 17, Fig. 1)

Hazara *paludar:* Kashm. *badar, drewar:* Himachal *Pindrau, tos, span, rau;* Garhwal *morinda;* Kumaon *ragha.*

Etymology : The specific name is after the vernacular name 'pindrau' prevalent in the Simla hills.

Description : A lofty evergreen tree with a dense narrowly cylindrical crown (see photograph) of very dark green foliage. Height mostly 45-60 m girth, 2.5-4 m, upto 63 m and girths of 6 m have been recorded. *Bark* smooth on young stems, dark-grey, rough with deep vertical furrows on old stems. *Leaves* variable, 2.5-6.2 cm long, some times longer two ranked, on opposite sides of a twig; needle-like, flattened, undersurface with 2 silvery bands on either side of midrib, tip notched. *Male and female strobili* on the same tree. Male catkins 1.3-1.8 cm long, clustered; stamens with 2 linear pollen sacs. Female strobili in cones which are solitary or in distant pair, erect, dark purple, perfectly cylindrical 10-18 × 4-7.5 cm when mature. Scales 2.5-3 cm long by 3-3.8 cm broad, fan-shaped, prominently eared bearing two winged *seeds* 2.5-3.2 cm long including the wing and 1.2-2 cm long without it, brown shining; wing 1.2-3 cm wide. Cones break up on the tree.

Distribution : Afghanistan to Nepal throughout the Western Himalaya at 2300-3350 m sometimes descending below 2,150 m. It is mostly associated with deodar, spruce *Quercus dilatata* and *Q. semecarpifolia*. The last two are not met with in the Kashmir valley proper.

Phenology : The new needles appear in April-May. They persist for 3-6 years. The old needles fall chiefly in May-June. Male and female strobili appear in April. Cones ripen in September and break up on the tree shedding the seed in Oct.-Nov.

Uses : The timber is used in making fruit cases, planking for ceilings, floor boards and shingles, for houses in the Himalaya.

Nomenclature : According to Sprague (1933), the correct name of the Pindrow fir is *A. pindrow* Royle, not *A. pindrow* Spach, as cited in many works of reference including Dallimore & Jackson (1966). Spach himself (Hist. Nat. Veg. Phan II, 423. 1842) correctly cited *Abies pindrow* from Royle Ill.t.86. 1836. The figure of *A. pindrow* Royle, is accompanied by analyses and the name was, therefore, validly published in 1836. The principal synonymy of *A. pindrow* follows : *A. pindrow* Royle Ill.t.86, Fig. a-c (May 1836) et. pp 350, 351 (Feb. 1839). *Pinus pindrow* D. Don in Lambert Pinetum, ed. 1837-42, 3, 47-48, t 99. (1837).

Picea pindrow (D. Don) ex Loud. Arb. et. Fruct. Brit. IV, p. 2346 (1838); et in Royle Ill. Bot. Himal. P. 354 (Feb.) 1839.

A. webbiana var. *pindrow* Brandis For. Fl. 528, 1874.

A. pindrow Royle var *brevifolia* Dallimore et Jackson (*A. gamblei* Hickel) has been collected from Garhwl. It is characterized by shorter, rigid, acute, paler green leaves and brown branchlets not grey. It differs from *A. spectabilis* in its glabrous shoots, comparatively shorter and narrower leaves.

2. **A. spectabilis** (D. Don) Spach, Hist. Nat. Veg. Phan 11, p. 422, 1842. Raizada & Sahni 125-126, 1960. Dallimore & Jackson Handb. Conif. Ginkgo, rev. Harrison. ed 4. 84. f. 18, 1966. — **HIGH LEVEL FIR** (Pl. 17, Fig. 2)

Pinus spectabilis D. Don in Lamb. Descr. Gen. Pinus 2.3.t.2. 1824. et Prod Fl. Nepal 55, 1825. *Abies webbiana* (Wall) Lindl. Penny. Cyclop. 7. 1833. Hook.f. FBI 5, 654. 1890. Brandis Ind. Trees 692. 720. 1906.

Gobra Salla Nepal, Hind, *Bang, Chilrao, ragha ransula.* Kashmir *Badar, paludur.* Sikkim Bhutan: ***Dhunsing.***

Type : *Pinus spectabilis* (Parker p. 684, 1927).

Etymology : One of its names, *A. webbiana* in vogue earlier is after Capt. W.S. Webb who discovered it in SE Garhwal on one of the spurs of Trisul.

Description : A tree resembling *A. pindrow* but smaller, sometimes gnarled in the W. Himalaya with a wider crown, not cylindrical, somewhat pyramidal. *Bark* dark grey, deeply longitudinally grooved and scaling on old trees. *Branches* slightly up turned, upper branchlets horizontal and flattened. *Leaves* 1.3-2.5 cm long (W Himalaya), up to 6.2 cm (E. Himalaya), distinctly parted on the upper side of the branchlets and point sideways, apex of leaf rounded and notched; leaves flattened, pale beneath with incurved margins. The leaves are therefore much shorter than *A. pindrow.* Distinguished by the young branchlets which are hairy in the grooves but soon become hairless. *Female cones* dark purple, erect, nearly cylindrical *c* 7 cm long (much longer in cultivated specimens, 10 to rarely 20 cm.). They are therefore distinctly shorter and thicker than in *A. pindrow.* The bracteole is more rounded with thicker and longer mucro or apex than in the Pindrow fir. Male catkins 0.8-1.2 cm long.

Distribution : All along the W & E Himalaya including Tibet (Jingwei 1982). Afghanistan, Burma, 2800-4000 m.

Specimens examined : GARHWAL, *K.C. Sahni* 24120, Harki Doon, 11,700′; *K.C. Sahni* 27007, near Banderpunch, *J.S. Gamble* 24372, Harki Doon, annotation, small leaves. Type : *J.S. Gamble* s.n. Harki Doon 13000′: DARJEELING, Singlila Range 10500′, thru Wood Technologist FRI, No. 1. BURMA, *C.W.D. Kermode* 17267, Myitkyina 9000′ Fide Kew.

According to Parker, p. 690 (1927) '*A. pindrow* and *A. spectabilis* are scarcely ever found growing together. *A spectabilis* commences 500-1000′ above the highest *A. pindrow*'. In the W. Himalaya it occurs towards the upper limit of tree growth associated with *Betula utilis* D. Don and *Rhododendron campanulatum* D. Don, rarely ascending to 4250 m.

According to Thierot (1980), p. 520, 'A fir (* *Abies spectabilis*) has been found at 5350 m/c 17,500′ in Kashmir.

Parker op. cit. p. 684 was of the opinion that *A. spectabilis* does not occur in the E. Himalaya. The author is not in agreement with Parker on this point. In the Eastern Himalaya in Sikkim and in Arunachal Pradesh where the author made collections and carried out field surveys it is a much larger and taller tree up to 60 m with longer needles up to 6 cm, it grows below the zone of *A. densa.* In the W. Himalaya it grows above the zone of *A. pindrow.* Regarding the W. Himalaya high-level fir (*A. spectabilis*) Gamble (1902) stated "I believe that this species is an East Himalayan one which extends westwards at high elevation..."

Uses : Timber used for building and roof-shingles.

Birds : The Monal Pheasant is a dweller of fir, birch and *Rhododendron campanalatum* forests.

3. **Abies densa** W. Griff. ex Parker in Ind. Forester 53 (12): 686. (1927) Hara, Stearn & Williams En. Fl. Pls. Nepal. Vo. I: 25, 1978. *Abies densa* Griff., J. Trav. : 258 (1847) nom. nud. Mast. in Curtis' Bot. Mag. 132: t 8098, 1906. Brandis Ind. Trees, 692, 1906.

E. HIMALAYAN FIR, RED FIR (Pl. 14, Fig. 2)

Gobra Salla, Nepal : *Dunshing* - Bhutan.

Etymology : Densa from the crowded leaves.

Description : Tree with an umbrella-shaped or pagoda-shaped crown with horizontally long spreading flattened branches, fissured. Young shoots hairy, reddish brown (hence the popular name Red Fir). *Leaves* crowded, more or less distichous, needle-like, flattened, 1.3-5 cm long, under surface silvery glaucous, margin not revolute. *Male and Female Strobili* on the same tree. Male catkins 4-5 cm, cylindric. Stamens with 2 linear pollen sacs. *Female cones* erect, scales thin, 2.3 × 1.8 cm, breaking away from the woody axis when ripe, fan-shaped, contracted by a short claw. Ripe cones cylindric, thicker, and shorter than *A. pindrow* 10-17.5 cm long × 3.5-7.5 cm. *Seed* 0.5 cm, wings 2, 0.8 × 0.6 cm, seed with wing 1.3 cm long. Female cones in all Himalayan firs are a deep purple black.

---

* Specific name in parenthesis is author's. It cannot be any other fir in the N.W. Himalaya.

Distribution : E. Himalaya, E, Nepal, Sikkim, Bhutan, Arunachal Pradesh to SE Tibet at 2,760-3950 m, N.E. Frontier of Burma. Above 3200 m it from forests of considerable extent and is gregarious. Below this elevation it is sometimes mixed with spruce, and hemlock. At higher elevation it is stunted and is often associated with juniper and *Rhododendron campanulatum.*

Specimens examined : SIKKIM, *Arjun Singh* s.n; Dambang Reserve Forest, 10,000′; *Ribu & Rhomoo* 2962, 13000′, Dt. R.N. Parker; *K.C. Sahni* 605, Yumthang; *K.C. Sahni* Thangu (N. Sikkim), Tree 50′, 12,600′. DARJEELING, *J.H. Lace* 2295, Sandakphu 11,000′. ARUNACHAL PRADESH, *K.C. Sahni* 458 on Se La 11,000′, gregarious, Kameng Dt. *K.C. Sahni & H.B. Naithani* 735, Tawang, 10,000′; *H.B. Naithani* 1160, Jaswantgarh, near Jung, Kameng Dt, Det. Rushforth. BURMA, *Capt Abbey* s.n.; N.E. Frontier, Myitkyina Dist.

Some botanists consider this species to be a synonym of *A. spectabilis* (*A. webbiana*). I consider them distinct and the two are separated by the following characters.

(1) *A. densa* has a wide-spreading umbrella or pagoda-shaped crown (see photograph). (2) bigger cones 10-17 cm long, up to 7 cm long in *A. spectabilis.* (3) leaves 1.3-5 cm, crowded in *A. densa,* up to 6.2 cm in E. Himalayan trees of *A. spectabilis.* (4) *A. spectabilis* in the E. Himalaya is a much larger tree up to 150-200′, 20-30′ in girth and is found at lower altitudes than *A. densa.* (5) *A. densa* has a wider altitudinal range.

Sir Dietrich Brandis *loc. cit.* regarding this taxon says. 'The eastern form I have kept separate at the suggestion of Prof. Heinrich Mayr of Munich who visited the N.W. Himalaya and Sikkim in 1887. Mayr (Wald und Parkbaume fuer Europa p. 165, 174) saw the Sikkim fir and the low level fir (W. Himalaya) and noted the wide spreading, umbrella shaped crown of the former and the narrow cylindrical crown of the latter and concluded at once that they must be separated.

Mehra and Jain (1976) consider the W. Himalayan Fir as *A. spectabilis* and the E. Himalayan Fir as *A. forrestii* C.G. Rogers not *A. densa* Griff.

Uses : The same as for *A. pindrow.*

According to Hara, Stearn & Williams (1978, p. 25) 'This Silver Fir (*A. densa*) has been for a long time confused with *A. spectabilis,* but is well distinguished by the bark soon becoming scaly, twigs brownish (not yellowish) leaves broader (2.3-3.3 mm wide), mostly revolute on the margins (at least on dried specimens), those of twig-underside mostly falcate and spreading laterally in two ranks; male flowers 4-5 cm long, cone scales 14-22 × 15-22 mm, thick, the protruding bracts widening distally in a spathulate blade 6-9 mm wide and ending in a cusp 2-4 mm. It is a tall tree to 60 m, with long spreading branches forming a broad flattened crown', The above description by Hara *et al* has been given in full as till today there is confusion regarding *A. densa* and *A. spectabilis.* O. Polunin & Stainton (1984, p. 386) also consider *A. densa* distinct from *A. spectabilis.* They have said 'Very like 1308 (*A. spectabilis*), and often confused with it, but distinguished by its bark which soon becomes scaly, by its brownish (not yellowish twigs), and by its leaves mostly with incurved margins. Cones with bracts which protrude between the scales and widen into a spathulate blade ending in a long point'.

4. **Abies delavayi** (Van Tiegham) Franchet in Journ. de Bot. 13: 255, 1899. Raizada & Sahni 128, 1960. Dallimore & Jackson rev. Harrison. Handb. Conif. and Ginkgo ed 4. 51, 1966. — **DELAVAY'S FIR** (Pl. 18, Fig. 1)

A. fabri (Masters) Craib.

Etymology : After Abbe Delavay explorer botanist who botanised in China.

A predominantly Chinese species so far found only in 3 localities in India. First by Bor (1935) from Peri La and later by Sahni from Peri La and above Tawang in 1978 (Sahni 1978) and Se La. These 3 localities are in Kameng District. Its collection presents difficulties as Peri La in Kameng District in enshrouded in mist almost all the year round. The most distinctive feature of this fir is in its needles which have a strongly revolute margin a character not shared by any other Chinese or Indian firs.

Description : A tree 18-40 m/130′ high with massive branches when old. *Young shoots* reddish-brown or yellowish brown, grooves slight or absent, minutely pubescent. *Winter buds* ovoid 0.3 cm long, reddish brown, resinous. *Leaves* on lateral branchlets arranged in two opposite sets, those of the upper side of the shoot much shorter than the lower leaves which are directed outward and downwards. 1.8-3.1 cm long, rigid, dark, glossy green and furrowed above; midrib prominent beneath with 2 broad whitish bands of stomata, margins usually strongly revolute so as to cover the stomatic bands beneath. *Cones* 6.27-10 cm long broadly cylindrical or barrel shaped, dark purplish-black, attractive, the apex of the bract slightly exserted when the cone is ripe.

Distribution : S.W. Szechuen, China, N.E. Upper Burma (collector Forrest. see Orr, M.Y. Notes R.B.G. Edin 18 (LXXXVIII) ; 142, 143, Nov. 1933. Bhutan, Kameng District, Arunachal Pradesh. In Arunachal Pradesh it forms stands between 2750-3250 m on northern slopes and has been collected by the author from two localities in Kameng viz Peri La and above Tawang, and from Se La by R.N. Loganey No 15. and determined, by the author.

Specimens collected/examined : ARUNACHAL PRADESH : Peri La 10,000′, *K.C. Sahni & H.B. Naithani* 826; Above Tawang gregarious ♀ cone common at 3000 m and above *K.C. Sahni & H.B. Naithani* 745 and 756; on way to Bumla from Tawang, *R.N. Loganey* No. 18; Se La 14000′, *R.N. Loganey* No. 15. Det. K.C. Sahni - 'strongly revolute margin' -annotation by author. BURMA, Htama 10,000′. *Tha Hla & Chit Ko Ko* 4147. BHUTAN, Kyimpa 12,500′ *Ludlow & Sherrif* 1572 (E) - Fletcher p. 91, 1976. Specimen not seen by author.

According to Dallimore & Jackson (revision), Harrison op. cit. p. 52 '*A. delavayi* has a greater altitudinal and latitudinal range than any other Chinese species and shows a considerable degree of variation. Its varieties were formerly regarded as distinct species, but it has become evident that they were too closely allied to be given a rank above varieties'.

Phenology : Fr. May. Leaf falling period is not evident (Academia Sinica Peking) ♀ cones mature Oct. Nov. (Sahni 756).

The timber is valued on account of the large size of the logs. *A. fabri* (Masters) Craib was considered distinct from *A. delavayi* mainly on the grounds that its leaf margins were only slightly revolute, its branchlets puberulent, and its cone scales reflexed. According to Dallimore & Jackson ed 4, p. 51, 1966 it seems that these characters are not consistently different, so Craib's species must be relegated to synonmy'.

Dallimore and Jackson op. cit. p. 53 mention a variety; *A. delavayi* Franchet var *forrestii* (C.C. Rogers) A.B. Jackson. 'It is characterised by needles with slightly recurved margin, found in Yunnan, where it forms forests at 10,000-11,000′, also in Szechwan'. Mehra & Jain (1976) on the basis of their morpho-anatomical studies conclude 'The East Himalayan Fir is identified as *Abies forestii* C.C. Rogers, and not *A. spectabilis* (D. Don) Spach, or *A. densa* Griff .....'

In Kameng District, *A. delavayi* and *A. densa* grow in the same areas, at almost similar altitudes and therefore present difficulties in quick identification as their crowns are of almost of the same shape.

The author separates the two by the following characters :

Needles relatively short 1.8-3.1 cm and rigid with strongly revolute margins in mature and immature needles, prominent mid rib. ♀ cones 6.2-10 cm, barrel shaped, purplish black, bracts exserted, recurved	*A. delavayi*
Needles longer 1.3-5 cm with flat margins when mature or only slightly revolute when immature (or in dried herbarium specimens). ♀ cones 10-17.5 × 3.5-7.5 cm, cylindric	*A. densa*

5. **Cedrus** Trew *nom cons.* Trew, *Cedrorum Libani* Hist. 1:6. 1757. — **TRUE CEDARS**

Etymology : Cedrus is taken from the ancient Greek, Kedros, the name for a coniferous tree.

Evergreen trees distributed in the Syrian Mountains, Atlas Mountains, Cyprus and the Himalaya. Pyramidal shaped when young, often developing massive trunks and large table-like branch systems when old, the foliage is borne in tufted masses except on terminal shoots.. *Bark* of young trees greyish, thin and smooth, on mature trees, brown, thick, deeply furrowed and broken into small irregular plates. *Leaves* needle-like, persistent, lasting 3-6 years, usually 3-sided. *Male* and *female* strobili on the same or separate trees. Male strobili solitary, at the ends of short shoots, catkin-like, but erect, stiff, consisting of numerous spirally arranged stamens, each with two adjacent pollen sacs opening longitudinally; connective projecting an expanded appendage, often egg shaped, toothed; pollen golden without wings. Female strobili in small greenish cones solitary at the tips of side

shoots; ovuliferous scales numerous in many rows spirally arranged, rounded externally and narrowed to a claw like process at the base; bract scales often enlarging after flowering and longer than the fertile scales. *Cones* large erect, barrel shaped, seated on a stout stalk. ·Two *seeds* are found on each scale, attached to a large membranous apical wing several times larger than the seed. Cotyledons 9-10. The female cones are fully mature two years after pollination and break up while on the branches. The last character separates the genus from *Larix* which it superficially resembles; in *Larix* the cones do not break up. In *Cedrus* the cones are cylindrical, barrel shaped. In *Larix* the cones are pendulous much narrower in comparison to their length. In *Cedrus* the leaves are evergreen, in *Larix* deciduous. The subcontinent has only one species *C. deodara* (Roxb.) G. Don in the Himalaya. Outside the Himalaya there are 3 species, *C. libani* (Lebanon), *C. altantica* (Algeria) and *C. brevifolia* (Cyprus). Some botanists consider them to be geographical sub-species or varieties of *C. libani.*

1. **Cedrus deodara** (Roxb.) G. Don in Loud-Hort. Brit. 388. 1830. Raizada & Sahni 129, 1960; Dallimore and Jackson. Handb.-Conif. and Ginkgo-rev. Harrison ed. 4. 142. 1966. *Pinus deodara* Roxb. Hort. Bengal. 69. 1818 nom. nud. et Fl. Ind. ed 2.3: 651, 1832.

DEODAR, HIMALAYAN CEDAR (Pl. 13, Fig. 1)

Cedrus libani Rich. var. *deodara* (Roxb) Hook.f. Himal. Journ. 1: 257, 1854. et FBI 5: 653. 1888.

Devadaru, Deodaru Sans : *Deodar* Hindi.

Etymology : The species name is derived from the Sanskrit *devadru,* tree of the gods (whence the common Indian name deodar).

Description : An evergreen tree exceeding 60.9 m in height and 10.6 m in girth. *Bark* greyish-brown. Leading shoots of young trees and branches pendulous. Branches of old trees horizontal, table-like. *Leaves* 2.5-3.8 cm long dark green, sharply pointed, in tufts of 15-20. *Male and female strobili* often on separate trees, but sometimes on the same tree in which case they are on separate branches. *Cones* erect, barrel shaped, 7.6-12.6 × 5.8-8 cm. Scales numerous, fan shaped, 3.2-3.8 cm across. *Seeds* triangular with a broad wing, the seed and wing 2.5-3.2 cm long.

Distribution : From Afghanistan to Garhwal, Kurnauli valley W. Nepal at 1200-3300 m. Commonly between 1700 to 2400 m.

Phenology : New shoots appear in March or early April. The old leaves are shed, chiefly in May, but shedding may also take place in autumn about the time the cones ripen. *Male* strobili appear in June. They ripen and shed their pollen from Sept.-Oct. and become yellow with pollen. *Female strobili* appear in August and ripen after pollination in November of the second year i.e. flowering to ripening is about 13 months.

Age/Dimensions : The upper reaches of the Ganga Valley in Garhwal have some fine forests of deodar. In a mature forest, trees from 2.4-3.6 m in girth predominate though far larger trees occasionally up to 76 m in height and 13.7 m girth are found. The largest deodars are usually seen near temples where they are venerated and protected as in Himachal Pradesh. In the Museum of the Forest Research Institute, Dehra Dun, there is a 2.5 m wide section of Deodar more than 750 years old, dating back to the 13th century when Qutb Minar was built. There are some immense stands of Deodar in Manali and in Garhwal. The Qutb Minar is 73 m and the tallest deodar of 76 m would over tower it.

Uses : It is the strongest of Indian coniferous woods. owing to the presence of oil, it is durable and rarely attacked by termites.

Birds in deodar forests. The Western Tragapon a most spectacular pheasant once occured from the Pakistan Himalaya through Kashmir to Kumaon. Now endangered it is now possibly localised in Kashmir and Himachal Pradesh.

6. **Larix** Miller Mill. Gard. Dict. 2:2. 1754. **LARCHES**

Deciduous trees distributed in the colder regions of the northern hemisphere, in Europe, Asia from the Eastern Himalaya northwards and in N. America. There are 10-12 species. *Leaves* are arranged as in *Cedrus,* but soft, light green and deciduous. *Male* strobili globose to oblong. *Cones* sub-globose, ripening in one year. Scales few or many, persistent. Bracts narrow, often bright coloured, longer than the scales at first, at maturity longer or shorter. *Seeds* small, winged cotyledons usually 6.

Larix resembles *Cedrus* in branch and leaf arrangement. In *Cedrus* the leaves are evergreen, dark green and the cones break up on the branches. In

Larix the leaves are deciduous, light green and the cones hang for a long time on the branches. Larch is the only deciduous conifer in the subcontinent with two species, *L. griffithiana* in the Eastern Himalaya with a wide distribution Nepal-Arunachal Pradesh, Tibet and Upper Burma and *L. himalaica* endemic to C. Nepal and S. Tibet.

Key to Larix

Female cones 5-10 cm long, dark brown when ripe.................. 1. *L. griffithiana*

Female cones 4-5 cm long, light brown when ripe.................... 2. *L. himalaica*

1. **Larix griffithiana** (Lindl. et Gord.) Hort ex Carr. Traite Conif. 278, June 1855. Raizada & Sahni 132, 1960. *Abies griffithiana* Hk.f. ex Lindley et Gordon in J. Hort. Soc. Lond. 5: 214, 1850, *nomen sub nudum.* **HIMALAYAN LARCH** (Pl. 13, Fig. 3)

Larix griffithii Hk.f. et Th. ex Hk.f. Illus. Him. Pl. t. 21, Excl. Fig. 1-4, July 1855; Hk.f. FBI 5: 655. 1883; Dallimore and Jackson. Handb. Conif. and Ginkgo. rev. Harrison ed 4, 303, 1966.

Pinus griffithii (Hk.f. et Thoms.) Parl. in DC. Prodr. 16(2): 411, 1868.

Pinus griffithiana (Lindl. et. Gord.) Voss. in Mitt. Deutsch. Dendr. Ges 16: 93. 1907.

Boarga Sella, Binya Nepal; *Sah, Saar* Lepcha.

Description : A graceful tree up to 18 m with long (weeping) pendulous branchlets. *Bark* thick brown. Young shoots downy, reddish-brown the second year. Terminal buds conical, covered with hairy scales; lateral buds egg-shaped, downy. *Leaves* 2.5-3.2 cm long, light green, ending in a black point, deeply keeled, below in tufts of 10 to about 19, deciduous. *Male strobili* 0.9 cm long on short, stout stalks. Female strobili ovoid, bracts long reflexed. *Cones* cylindrical 5-10 × 2.5-3.2 cm, coppery red during growth turning dark brown when mature, the short stout stalk twisted to bring the point of the cone upward on the pendent shoot. *Seed* 0.4 cm long, winged, in pairs on each scale 1-1.6 cm long including the wing. Wing about 1.2 cm long.

Distribution : Eastern Himalaya from E. Nepal through Darjeeling, Sikkim, Bhutan, Arunachal Pradesh, N E Upper Burma, and Chumbi Valley in Tibet at 2400-3650 m mostly on glacial moraine.

Specimens examined : NEPAL : Central, *Stainton* 6682; East : *Stainton* 1123. 1400-3900 m. (*BM)

Phenology : The needles are shed in the autumn, the new foliage appearing in the spring. The strobili appear in May and the cones ripen in October of the same year. A very graceful ornamental looking tree particularly when the cones are developing which are coppery red of the colour of 'Litchi' fruits. In Sikkim the wood is used for making shingles.

Nomenclature : The name *L. griffithiana* Hort. ex Carriere is to be preferred over *L. griffithii* Hk.f. et Th. According to Stearn (1938), both names were first validly published in 1855. Both have been adopted in works of authority. Since Carriere's Traite had been offered for sale in Paris and had reached England while Hooker's Illustrations was still in the printer's hand *Larix griffithiana* Carriere has clear priority over *L. griffithii* Hk.f. et Thomson. Its synonmy is given at the beginning of the account.

2. **Larix himalaica** W.C. Cheng & L.K. Fu in Acta Phytotax. Sinica 13: 84, f. 26, 1-6 (1975). A tree, distinguished from *L. griffithiana* by its yellowish-grey branchlets, shorter and broader *Cones* 4-5 cm long, which are light brown when ripe, and with broad, minutely irregularly toothed cone-scales and erect bracts about as long as the scales and abruptly narrowed into rigid tooth. *Leaves* to 3 cm. (Polunin & Stainton 1984).

NEPAL LARCH
TIBETAN LARCH
(Pl. 19, Fig. 1)

Distribution : Endemic to C. Nepal and S. Tibet at 2400-4000 m in mixed forests. C. NEPAL : *Gardner* 863; *Nicolson* 2558 : *Polunin* 536; *Stainton* 3916, 2400-3600 m (•BM)

According to Hara, Stearn & Williams (1978), closely akin to *L. potaninii* Batalin from W. China (Yunnan, etc), but distinguished by

* Specimens not examined by me.

• Specimens not seen by author. Not in DD.

yellowish grey twigs, larger cones, light brown when ripe, with minutely erose scales and bracts truncate and contracted into a cusp.

Mehra & Jain (1976) have contributed on leaf anatomy and wood structure of *Larix.*

Family 8. **TAXODIACE** Warming *nom. cons.*

Warming, Handb. Syst. Bot. 184. 1890.

RED WOOD FAMILY TAXOIDS

Large trees with evergreen or deciduous *leaves* which are scale-like or needle or sickle-shaped. *Monoecious;* the males either terminal or lateral forming heads or catkins; stamens with a short stalk with 2-8 pollen sacs hanging from the undersurface of the broadly expanded connective, pollen without wings. Female *strobili* solitary, terminal, surrounded at the base by a few scales and consisting of spirally arranged scales of two sorts; fertile scales more or less adnate to the bracts, during flowering time present as cushion-like swelling, later on getting larger and overlapping the bracts, sometimes absent or inseparable from the bracts (*Taiwania*) : ovules two, upright or inverted. *Cone* globose or ovoid, with numerous overlapping, rounded, mucronate scales, each with a minute bract at the base. *Seed* with a woody shell shortly winged.

Ten genera with 16 species distributed in E. Asia, Tasmania and N. America. *Taiwania cryptomerioides* Hayata is the only genus in our area found in Burma. *Cryptomeria* and *Taxodium* are exotics which have been introduced in India, the former on a plantation scale in the Darjeeling Hills and the latter in Dehra Dun.

These 3 are separated by the following Key.

Side shoots leafy, deciduous as a whole.............................. 1. *Taxodium*

Leaves and shoots not deciduous

  Fertile scale and bract distinct.................................. 2. *Cryptomeria*

  Fertile scale and bract not distinct.............................. 3. *Taiwania*

Out of the Redwoods only the Dawn Redwood (*Metasequoia glyptostroboides*) has established in India limited to a few trees in Darjeeling. It survived for 5 years at the FRI, Dehra Dun.

As suggested earlier it should be tried along a water course at the FRI. It was discovered as a fossil in Japan by Miki in 1941, later a living tree was discovered by Chinese botanists and foresters in C. China where it grows in shady moist places at 700-1350 m. It is monotypic. Outside China it failed to produce fertile pollen. A fast grower upto one meter a year, it has a forestry potential and is propagated by cuttings struck during summer. (Florin 1963, Hora 1981).

Sequoia sempervirens (Coast Redwood) was tried in Chakrata in *c* 1952, a living sapling was brought from the U.S.A. by Sri K.M. Munshi Minister for Forests and survived only for a year. Its native habitat is sea level-900 m in the foggy California coast from Oregon to Monterey on seaward side of mountains on banks of streams. It is worth a trial in the mist enshrouded Darjeeling hills at *C.* 1500 m. Its maximum height is 120 m/ 400′ and is probably the tallest in the world. The Sierra Redwood (*Sequoiadendron giganteum*) is confined to Sierra Nevada at 900-2700 m in cooler drier climate. For its trial, matching climates are to be located in the Himalaya. It is most essential to find out from American forestry literature whether it occurs in a winter rainfall zone or summer rainfall zone of N. America. If it occurs in a winter rainfall area then Kashmir Valley may be more suitable than other parts of the Himalaya which fall in the monsoon belt. *Cunninghamia lanceolata* (Chinese Fir) has been established in Dehra Dun and is briefly described towards the end of the account of this family.

1. **Taxodium** Rich. in Ann. Mus. Hist. Nat. Paris 14: 290. 1810. **BALD CYPRESS**

Deciduous or semi-evergreen trees with light brown scaly *bark.* *Branches* erect or spreading; branchlets of two kinds, persistent with axillary buds near the end of the year's growth, and decidous without axillary buds, falling during the autumn. The *leaves* are spirally arranged on the branches, but on the deciduous side-shoots the leaves appear two-ranked owing to a twist near the base. *Flowers* monoecious. Male flowers in purple, slender tassel-like clusters, stalked, scaly at the base. One species an exotic *T. distichum* does well in cool damp climate and has done well in Dehra Dun. Some magnificent trees are seen along the canal in the Forest Research Institute, Dehra Dun.

T. distichum (L) Rich. in Ann. Mus. Hist. Nat. Paris: 14:298, 1810. Raizada and Sahni 132, 1960. Dallimore and Jackson. rev. Harrison. Handb. Conif. Ginkgo ed 4. 534. 1966. *Cupressus disticha* Linn. Sp. Pl. 1003, 1753.

BALD CYPRESS, MARSH CYPRESS
(Pl. 21, Fig. 4)

Etymology : From the Greek taxos, yew tree, and edios, resemblance, from the leaves, which resemble those of other species of Taxus.

Description : A deciduous tree upto *c* 40 m/130′ and 1.2-1.8 m in girth. The reddish-brown *bark* has long shallow cracks. In swampy places the trunk develops enlarged, fluted, buttress-like roots and protuberances (knees) which take air from the roots. Branchlets at the point of shoots persistent, lateral branchlets falling with leaves in autumn. *Buds* near the tips of shoot rounded with overlapping sharp pointed scales. *Leaves* spirally arranged, linear pointed *c* 1 cm long, light green, in the autumn, they change from tawny to brown. Male and female *strobili* appearing on the same tree in spring, the former in slender, purplish, tassel-like clusters 7.5-15 cm long, (rarely 18.5-27.5 cm). Male strobili stalked consisting of 6-8 stamens and surrounded at the base by ovate scales. Female strobili scattered, solitary or several together, consisting of numerous overlapping pointed bracts adhering below to the fleshy scales, each of which bears 2 ovules. *Cones* globose or obovoid, resinous, 1.2-1.3 cm in diameter, purplish, on stalk 0.2 cm long, scales few, 4-sided, narrowing to a long slender stipe. *Seeds* irregularly triangular, with horny, warted coats.

Uses : It is a majestic and beautiful tree suitable for parks and for planting along water courses. It has done extremely well in Dehra Dun along canals in the Forest Research Institute and also in big compounds on Rajpur Road, Dehra Dun away from water courses where water supply is limited but there is abundant rainfall. This shows that it is fairly adaptable under Dehra Dun conditions. It is very fast growing along water courses. Valued for its resistance to decay and termites.

Propagation : By seed, or by cuttings in ground saturated with water.

2. **Cryptomeria** D. Don in Trans. Linn. Soc. Lond. 18: 166. 1841

JAPANESE RED CEDAR

A monotypic genus of China and Japan introduced on a plantation scale in Darjeeling. Tall trees with drooping branches. *Leaves* spirally arranged, awl-shaped 0.63-1.27 cm. Keeled on both surfaces. Monoecious. Male subterminal in clusters of 20 or more, stamens spiral, subsessile, each with 4-5 round anthers. Female terminal, consisting of 20-30 spirally arranged scales and bracts; ovules 2-5, upright. Female cone 2-3 cm with 20-30 wedge shaped scales. *Seeds* dark brown, irregularly triangular 1.25 cm long × 0.6 cm wide, each edge narrowed into a rudimentary wing.

It exists according to Bayard Hora (1981) into 2 geographically isolated varieties var. *japonica* from Japan and var. *sinensis* native to China.

Cryptomeria japonica (Linn.f.) D. Don in Trans. Linn. Soc. London 18: 167 t 13.f 1. 1841. Dallimore and Jackson Handb. Conif. ed. 3. 252. f. 42. 1948. Raizada and Sahni 135, 1960. **JAPANESE CEDAR** (Pl. 21, Fig. 3)

Etymology : The name is derived from the Greek *cryptos,* hidden, and meros, part, because the parts of the flower are not easy to distinguish.

Habitat : Native to Japan, forming vast forests (var *japonica*), and to south China var *sinensis* introduced in the Darjeeling Hills on a large scale.

Description : A tree up to 50 m/165′ with a pyramidal crown and straight trunk with red-brown fibrous *bark* pealing off in long strips. *Needles* awl-shaped, 1-2 cm, arranged spirally, persisting for 5 years. Monoecious. *Male strobili* clustered at the end of the branchlets, *female* at the end of the branchlets, globose cone-like, covered with small awl-shaped leaves. The spherical *cones* 1.2-1.9 cm are terminal; they ripen in the first year and remain on the tree for many months after dissemination has taken place. *Seeds* dark brown, irregularly triangular 0.65 cm × 0.3, each edge narrowed to a rudimentary wing.

Uses : The Darjeeling wood is inferior to that of Japan. It is used for making light packing cases. With suitable chemical treatment bottle corks can be made from it. (Narayanamurti et *et al.* 1941).

3. **Taiwania** Hayata in Journ. Linn. Soc. Bot. 37: 330. 1906.

A monotypic genus of evergreen trees, native to Taiwan, China, Manchuria and North Burma. Male strobili in groups of 5-7, collected at the ends of short side shoots, each consisting of about 15 stamens standing at right angles to the axis; filament filiform; connective expanded at the tip into an oval toothed scale bearing at its base two free anthers. Female strobili in shortly cylindrical *cones* each consisting of 12-20 scales, spirally arranged, leathery, each with two ovules; bract not distinguished from the scale. *Seeds* 2 to each scale.

Taiwania cryptomerioides Hayata J. Linn. Soc. (1906) 37, 330, pl 16. Raizada and Sahni 136 1960. Dallimore & Jackson. Handb. Conif. and Ginkgo. rev. Harrison. edn. 1966, 582. *T. flousiana* Gaussen. **CHINESE COFFIN TREE** (Pl. 20, Fig. 2)

Etymology : The generic name is derived from the island of Taiwan (Formosa) where it was discovered in 1904, on the slopes of Mt. Morrison at 6,000-8000′. The specific name is derived from *Cryptomeria* as in foliage the two are alike.

The tree has also been found in Yunnan (China), and by Kingdon-Ward in N Burma (Botanical Exploration in N Burma Jl. R. hort. Soc., 71, (pt 2), 323.) Its occurrence in S.W. Yunnan and in the mountains of Taiwan and in no place in between is an interesting case of disjunct distribution. The continental and Taiwan provnences according to Dallimore and Jackson (1966) p 583 have been regarded as distinct species by some authors, but the characters used to distinguish them seem to be inconsistant and overlapping.

Description : An evergreen tree of conical form nearly 200′ high and 20′ or more in girth in Taiwan. *Trunk* bare of branches for about 60′ from the ground, with a spreading head and dense foliage. Adult *leaves* about 0.5 cm long, thick, leathery triangular, incurved at the shortly pointed apex, overlapping, pressed to the branchlets for about half their length; leaves of young trees narrowly linear, sickle-shaped, compressed, keeled above and below. Male *strobili* on some trees are extraordinarily abundant. *Cones* shortly cylindric subglobose 1.27 cm long, with numerous overlapping, rounded mucronate scales, each with a minute bract at the base. *Seeds* two on each scale, oblong broadly winged. According to Wilson (1926) fruit is found on the tops of the oldest trees. 2n = 22.

In foliage it is very much like *Cryptomeria* but in cone structure it has close affinities with *Cunninghamia* from which it, however differs in having 2 ovules instead of 3 to each fertile scale. Specimens collected many years ago from Burma were misindentified as *Juniperus recurva*. In 1939 the specimens were examined at the Herbarium of the Forest Research Institute,

Dehra Dun and were identified as *Taiwania cryptomerioides* and thus cleared up the mystery of the Chinese Coffin tree. However some Junipers, *Cunninghamia,* etc are also used as coffin woods. The more affluent prefer *Taiwania.*

Distribution : S.W. Yunnan (China), S.E. Tibet (Kingdon-Ward); watershed of the Salween and Irrawadi River (Burma, Handel Mazzeti specimens) and N. Burma (Kingdon-Ward).

Specimens examined : BURMA, *C.W.D. Kermode* 17: 59228.4.1938 Myitkyina, Kangtang village, 5500′; C.W.D. *Kermode* 17341, Kangtang 5000′. Tree 7-8′ girth.

Uses : In demand in China as coffin boards by wealthy Chinese. From N. Burma they are carried to China overland.

Cunninghamia lanceolata (Lambert) Hooker f. *Pinus lanceolata* Lambert; *Cunninghamia sinensis* Richard **CHINESE FIR** (Pl. 20, Fig. 1)

Etymology : Named after James Cunningham who discovered it in China in 1702.

This tree is an exotic from south and west China successfully introduced in the Forest Research Institute, Dehra Dun and at Chakrata. As it is not indigenous and not introduced on a plantation scale, it is very briefly described.

Description : Evergreen tree to 25 m/82′, much smaller at Dehra Dun 5 m (-8 m). *Leaves* stiff 3-6 cm × 2-6 mm. with 2 broad white stomatal lines. *Male and Female* cones occur on the same tree. Male oblong, terminal in clusters of *c* 12 at the shoot. ♀ cone 2.5-5 cm with 3 ovules per scale. *Fl. :* Formed in Oct.-open April-May *Fr. :* cold weather.

Uses : In China timber used for coffins. In view of its small size under Dehra Dun conditions it is suitable for small compounds. As it is attractive all the year round it is suitable for landscaping.

Family 9. **CUPRESSACEAE** Bartling

Bartling, Ord. Nat. Pl. 90. 95. 1830.

CYPRESS FAMILY

Evergreen trees/shrubs with scaly or needle-shaped opposite or whorled leaves. *Male/Female strobili* on the same or on different plants, terminal on short twigs or axillary; solitary and consisting of opposite or whorled stamens with a short filament; connective prolonged into an excentric peltate scale-like appendage; anthers 2-6 globose. Pollen grains not winged. Female strobili of one or two rowed, opposite or whorled scales, which are not at first separable into ovuliferous scale and bract, on a shortened axis; ovules few to many, sessile at the base of the scale, orthotropus. *'Fruit'* a cone with usually woody or fleshy scales. *Seeds* solitary or geminate, rarely numerous.

Juniperus and *Cupressus* are indigenous in the Himalaya. *Callitris* has been introduced in the Nilgiris on a plantation scale. *Thuja* an exotic from China is commonly planted in gardens in the subcontinent. *Calocedrus* was recorded from Burma where it is rare. This family is bihemispherical with 19 genera and 130 species. The genera *Juniperus, Cupressus* and *Calocedrus* are found in the N. Hemisphere, *Callitris, Libocedrus,* etc are confined to the southern Hemisphere.

Key to the Genera

Cones fleshy, berry-like .... 1. *Juniperus*

Cones dry of woody scales.

Cones of peltate, woody scales, valvate. Cones sub-globose.......... 2. *Cupressus*

Cones of one whorl of woody scales, the alternate scales shorter than the two on either side. Cones globular, ovoid or pyramidal.......... 3. *Callitris*

Cones with oblong scales which are thickened upwards. Cones oblong. 4. *Thuja*

Cones with fewer scales, woody. Cones linear oblong, narrowed at the apex........... .... 5. *Calocedrus*

The genera of *Cupressaceae* present some difficulty in identification in the field and in the herbarium. Keys based on cone characters as the one above are helpful. Juniper cones are fleshy and berry-like. *Cupressus* of woody peltate scales and cones of *Thuja* are usually oblong with scales thickened upwards or sometimes terminating in horn like structures as seen in *T. orientalis.* In *Calocedrus* the cones are oblong ellipsoid, scales 6, with a small dorsal woody spine. The leaf characters of these genera are not so helpful in identification. Juniper leaves are scale-like and needle-like, in *Cupressus* and *Thuja* they are very much alike, scale-like and decussate. The last two are quickly separated in the field. In *Thuja* the foliage is erect in *Cupressus* the foliage is pendulous. In *Callitris* the leaves are in whorls of three, those of the adult stage closely clasping the shoot except at the triangular scale-like tip.

1. **Juniperus** Linn. Sp. Pl. 1038. 1753. *Sabina* Spach. **JUNIPERS**

A genus comprising about 60 species of evergreen trees or shrubs distributed over the Northern hemisphere from the Arctic circle to Mexico and the W. Indies, Azores, Canary Islands, N. Africa, Ethiopia, the Sudan, the mountains of E. tropical Africa, Himalaya, China and Taiwan. The generic name *Sabina* has been used in connection with the junipers in some works but according to Dallimore & Jackson (1966) rev; Harrison p. 234 'this name is now not recognized'.

Bark thin peeling off in longitudinal strips. *Leaves* on young plants spreading and awl-shaped; on adult plants either awl-shaped and spreading or closely pressed and scale like. *Male and Female strobili* on the same or on separate trees, male strobili composed of numerous stamens on a central axis, with ovate or peltate scale-like connectives, each bearing 2-6 globose pollen sacs. Female strobili surrounded at the base by minute scale-like bracts, which persist unchanged under the fruit or cone, composed of 3-8 opposite or ternate, pointed, united scales, which become fleshy and form a berry-like fruit bearing one or more seeds. Wood fragrant and durable owing to the presence of oil, therefore used for making coffins. Seven species occur in our area. Mehra & Jain (1976) opine that *J. excelsa* Bieb of Greece occurs in the W. Himalaya. The author has not seen any wild trees in the field or in herbaria so far. Propagation is by seeds, cuttings or by grafting. According to Florin 1963 Junipers posses the capacity for long distance dispersal by birds.

Key to Juniperus

Trees

Leaves appressed.. 1. *J. polycarpos*

Leaves spreading.. 2(a) *J. recurva* var. *coxii*

Leaves scale like, decussate,.................................... 5. *J. indica*

Shrubs

Leaves needle like in threes .... 5. *J. indica*

Leaf bases not decurrent .... 3. *J. communis*

Leaf bases decurrent .... 4. *J. squamta*

Leaves scale-like or subulate and pungent.

Leaves lanceolate, in threes, pungent .... 2. *J. recurva*

Leaves scale-like decussate .... 5. *J. indica*

or

Leaves all needle shaped, in threes.

Leaf bases not decurrent .... 3. *J. communis*

Leaf bases decurrent. Under surface of the leaf furrowed from base upto halfway : no green midrib above .... 2. *J. recurva*

Under surface of the leaf furrowed from base to apex : no green midrib above .... 4. *J. squamata*

Leaves on old plants scale or only needle-like in young twigs. Twigs rounded .... 1. *J. polycarpos*

1. **Juniperus polycarpos** C. Koch *J. macropoda* Boiss Fl. Orient V, 709, 1884. Hook.f. FBI V, 647, 1888. Brandis Ind. Trees 694, 1906. Raizada and Sahni 138, 1960. Dallimore and Jackson. Handb. Conif.Ginkgo. rev. Harrison 261. 1966.

HIMALAYAN PENCIL JUNIPER (PL. 23, Fig.2)

Dhup, Padam Hindi; *Chalni, Lewar* Himachal *Obusht*-Pushtu; *Appurz, ghushki* Baluch; *Dhupi*- Nepal; W. Him. *Shir, Chalai, Dhup.*

Description : A shrub or medium sized evergreen tree 12-15 m or sometimes taller with a trunk of 1.8-2 m in girth. Trees up to 10 m girth and estimated to be 1000 years old are recorded from Lahul. *Leaves* of the lower branches linear *c* 8 mm pungent, of the upper branchlets scale-like *c* 1.5 mm closely and flatly pressed, egg shaped, pointed. *male catkins* on a scaly stalk. *Fruit* sub globose 8.5 mm across bluish-black when ripe, resinous 2-5 seeded.

Distribution : Afghanistan, Baluchistan, Kagan Valley, Kashmir, Lahul extending to Kumaon and W. Tibet from 2400-4,300 m in the inner semi-arid ranges. It forms open forests of considerable extent particularly in Kagan Valley. Gregarious with an underwood of *Prunus eburnea* Aitch. et. Hemsl., *Caragana gerardiana* Royle, etc.

It is the host of the parasite, *Arceuthobium oxycedri* M. Bieb. which causes considerable demage to trees which subsequently die.

Specimens examined : LAHUL 3000 m *U.C. Bhattacharya* 45180 ♀ cones; *U.C.B.* 45352. Dioecious tree. ♂ ; *U.C.B.* 40714, Keylong 3100 m, ♀ cones. All *B.S.D. Fl:* in spring. *Frs. :* ripen in Sept.-Oct. of the second year.

Uses : The wood is very suitable for furniture, buildings, pencil making and also burnt as incense in monastries.

2. **J. recurva** Buch-Ham ex D. Don Prodr. Fl. Nepal 55. 1825 Brandis Ind. Trees 694. 1906. Raizada and Sahni 139, 1960 Dallimore and Jackson Habdb. Conif. Ginkgo. rev. Harrison 269, 1966. W. Him. *Betar. guggal, thelu.* **DROOPING JUNIPER** (Pl. 23, Fig. 3)

A low spreading shrub or small tree of graceful drooping habit, somitimes to 10 m or taller. *bark* greyish-brown thin, peeling off in fibrous strips. *Branches* and branchlets recurved or pendent. *Leaves* awl-shaped, densely overlapping, 0.15-0.36 cm. long, ending in a sharp horny point; *male and female strobili* on the same or on different trees. Male strobili terminal or in the axils of the leaves of the ultimate divisions of the branchlets, stamens 12-16. *Fruit* axillary ripening in the second year, dark purplish brown to black, ovoid, 0.8-1.3 cm. *Seed* solitary, ovoid and pitted. A shrub in the W. Himalaya, a tree in the East Himalaya.

Distribution : In the inner ranges of the Himalaya at 2700-4600 m extending east to Arunachal Pradesh, Upper Burma, China and in the west through the W. Himalaya to Afghanistan. Gregarious often mixed with *J. communis.* A very wide ranging and variable species.

Specimens examined : KUMAON, Panch Chalhi glacier, 13800′ *K.C. Sahni* 20462 : ARUNACHAL PRADESH, Se La. 13000′. Kameng Dt., *K.C. Sahni* 467, above Tawang, *K.C. Sahni* 482 BHUTAN, Fletcher (1975) in 'A Quest of Flowers' **Ludlow & Sherrif* 1175 (E), Trees 45 × 10′ (girth).

Uses : Burnt as incense in monastries.

---

* Specimens not examined by me.

Juniperus recurva Buch.-Ham var. **coxii** (A.B. Jackson) Melville, Dallimore & Jackson Handb. Conif. Ginkgo. rev. Harrison edn 4, 269, 1966. ****COFFIN JUNIPER**

Sabina recurva (Buch.-Ham.) var. *coxii* (A.B. Jackson) et L.K.Fu in Fl. Reip . Pop. Sinicae, 7; 352, 1978.

Juniperus coxii A.B. Jackson in new Fl. and Silva 5: 33.f. 13. 14. 1932. Dallimore and Jackson Handb. Conif. ex. 3. 305. 322. 1948. Raizada and Sahni 139, 1960.

Etymology : named after E. Cox who botanised in E. Himalaya/Burma.

This new variety from N. Burma resembles *J. recurva* Buch.-Ham in its one-seeded fruit but differs in its longer and more acuminate leaves which are marked on the upper surface with 2 longitudinal greenish-white bands of stomata.

Description : A large tree. *Trunk* slender. *Branchlets* longer and more pendulous than in the type. *Bark* thick *Leaves* more laxly arranged, 0.6-0.8 cm long, long acuminate, the upper surface pale with a narrow, darker midrib. *Fruit* ovoid, 8 mm long, almost black, one *seeded* with a smooth and shining surface.

Distribution : Tsang Po Gorge, E. Himalaya.

Specimens of a Juniper 6314 collected by Kingdon Ward in 1924 from the Tsang Po Gorge in the E. Himalaya and preserved in the Kew Herbarium appear to be the same as var. *coxii* except that the leaves are somewhat shorter. It is described as a big tree 18-23 m in the temperate evergreen rain forest. Tsang Po Gorge has been described by Sir George Taylor FRS 'as the greatest existing sanctuary of horticultural and botanical riches...' (see Taylor p. XVIII in Fletcher 1975).

According to Dallimore & Jackson rev. Harrison op. cit. p. 270. "Well marked examples of typical *J. recurva* and the var *coxii* are very distinct, so much so that until recently the latter was regarded as a separate species. Dallimore suggested that *J. recurva* might well be regarded as the type of a widely distributed and variable species with the Coffin Juniper of Upper Burma as its most vigorous variant.....'

** A number of woods of *Juniperus* species are used as coffin wood because of their durability.

The generic name *Sabina* for *Juniperus* used by Cheng & Fu 1978 loc. cit is now not recognized.

3. **Juniperus communis** Linn. Sp. Pl. 1040, 1753. Hook.f. FBI V: 646. 1888. Brandis Ind. Trees 694, 1906. Raizada and Sahni 140, 1960. Dallimore and Jackson rev. Harrison Handb. Conif and Ginkgo 247, 1966.

COMMON JUNIPER GROUND JUNIPER
(Pl. 23, Fig. 1)

Betar, Bithal Chamba; *Pama, Tailu,* Kunawar.

Description : A dense erect shrub to 1.5 m. It has a wide distribution than any other shrub or tree, found throughout the temperate and subarctic regions of the Northern Hemisphere. *Bark* reddish brown scaling off in papery, sheets. Young shoots slender triangular with projecting ridges. *Buds* about 3.2 mm long with acuminate scales. *Leaves* awl-shaped, sharp pointed, sessile, spreading, 10-15 mm long, upper surface concave with a broad white band of stomata, lower surface bluntly keeled. *Male and female* strobili usually on different plants. *Male* strobili solitary. cylindrical, 8.5 mm long, the yellow stamens in 5-6 whorls. Female strobili solitary, green 2.1 mm long. *Fruit* ripening in the second or third year, green when young, bluish or black when ripe, globose or slightly longer than broad, 8.5-13 mm, diameter, with 3 minute points at the top. *Seeds* 2-3 elongated ovoid 3 cornered with depressions between.

Distribution : Afghanistan to C. Nepal 1800-3600 m. It is the most widely distributed shrub tree in the world. Temperate and subarctic Europe, Asia, N. Africa and N. America.

Uses : Oil of Juniper is distilled from the fully grown but unripe berries and is responsible for the characteristic flavour of gin.

4. **Juniperus squamata** Buch.-Ham. ex D.Don in Lambert, Genus Pinus, II: 17, 1824, Raizada & Sahni 140, 1960. Dallimore and Jackson rev. Harrison Handb. Conif. and Ginkgo. 275, 1966.

SINGLE SEED JUNIPER SCALY-LEAVED JUNIPER
(Pl. 24, Fig. 1-3)

Padma Chunder - Hindi.

A prostrate shrub in the Himalaya at 2400-4300 m associated with *J. communis* above the tree limit forming waist high thickets, the branches pressed down by snow. Main stems, decumbent and short, erect, reddish brown branchlets. *Young shoots* green, grooved, *leaves* awl-shaped, overlapping, in whorls of 3, pressed together or slightly spreading, the lower portion adherent to the shoot, the upper free and about 0.2 cm long, curved, tapering to a sharp point, upper surface concave, whitened, usually with a faint midrib; lower surface convex, green, furrowed. *Fruit* ellipsoid 0.6-0.8 cm long, becoming black when ripe in the second year; scales 3-6, pointed. *Seeds* solitary, ovoid, ridged, with 3-4 depressions below the middle.

Distribution : Afghanistan, Himalaya 2400-4300, usually associated with *J. communis,* N.E. upper Burma (Forrest collection), China, Taiwan and Japan.

Fl. : May-June. *Fr. :* June-Sept.

Uses : Used for fuel and burnt for incense.

5. **Juniperus indica** Bertol, Misc. Bot. 23: 228, t. 1 (1862); in Mem. Acad. Sci. Bologna ser. 2, 1: 228, t. 1 (1862) Hara *et al* Enum Fl. Pls Nepal, 27, 1978. *Juniperus wallichiana* Hk.f. et Th. ex Parl. in DC. Prod. 16(2): 482, 1868. Brandis Ind. Trees 695, 1906. Raizada and Sahni 141, 1960. Hara, Fl. E. Him. 42, 1966. Dallimore and Jackson Handb. Conif. Ginkgo. 4 edn. rev. Harrison, 284, 1966.

BLACK JUNIPER (Pl. 23, Fig. 4)

Juniperus pseudosabina auct. non. Fisch. & Mey Hook.f. FBI 5: 646 (1888).

Hindi *Bhil.*

Description : An evergreen tree upto 20 m in Sikkim and in the eastern Himalaya, with a stout trunk, becoming shrubby in the Western Himalaya up to one m. Adult *branchlets* with the ultimate divisions four-sided, the *leaves* closely pressed or overlapping, arranged in 4 ranks, narrowly ovate, 0.15 cm long, with the points incurved, furrowed on the back, margins whitish. *Leaves* on the main axis longer up to 0.6 cm long, with acuminate points; leaves awl-shaped in whorls of 3, 0.3-0.6 cm long, sharply pointed;

directed forwards, leaves on lower branches scale-like *c* 1.5 mm adpressed. *Male and female* strobili on different trees. *Fruit* ovoid 1.3 cm long, 0.8 cm wide near the base, brown, blue when ripe. *Seed* solitary 0.6 cm long, ovoid, compressed, with 2-3 depressions.

Distribution : All along the Himalaya and W. Tibet from the Indus to Bhutan at 2800-4600 m, associated with *Betula utilis* and *Rhododendron campanulatum.* In the Western Himalaya it is a shrub, in the eastern Himalaya owing to excessive humidity and rainfall it assumes the form of a tree up to 20 m. Joseph Hooker collected specimens in Sikkim and called it Black Juniper as the trunks are very dark almost black in colour. Not so far collected by the author or others from Kameng District, Arunachal Pradesh.

Fl. : April-May. *Fr. :* August.

Uses : Used as incense in monastries.

J. pseudosabina Fisch. et Mey has so far not been recorded from the Indo-Burmese region. It is a native of Altai and Tarabgatai Mts in Turkestan. It has leaves like *J. indica* but less acute. Mehra & Jain (1976) are of the view it occurs in the E. Himalaya at Jongri (W. Sikkim).

6. **Juniperus turkestanica** Komarov in Not. Syst. Herb. Hort. Bot. Reip. Ross. 5: 26. 1924: Raizada and Sahni 141, 1960. Dallimore and Jackson Handb. Conif. Ginkgo 279, 1966.

TURKESTAN JUNIPER (Photograph from KEW No. 3579 Pl. 25)

A tree up to 12 m/40′ high or a shrub less than 1.5 m with a dense crown. *Branches* erect or spreading. *Leaves* scale-like about 0.1 cm with strongly impressed glands. *Fruits* spherical or more frequently oblong 1.25-1.5 cm long, up to 1.25 cm in diameter, blackish, shining, somewhat glaucous, with very succulent flesh. *Seeds* oblong or spathulate, upto about 0.6 cm long and *c* 0.4 cm broad, grooved, with a hard testa. Some botanists placed it under *J. pseudosabina* Fisch. et Mey, which differs, however, in its much stouter branches, and cones twice the size with seeds of different character (Dallimore Jackson p. 279, 1966).

Distribution : Kazakihstan and Uzbekistan, in Central Asia. Pamir 1300-13,500′ (*H. Appleton* s.n., 1908, 6-10′ Kew Herbarium. Carl Troll

(1939) reported it from the Nanga Parbat Group. Mehra & Jain (1976) on the basis of morphoanatomical studies recognise the following taxa in the E. Himalaya, which are briefly summarized below.

1. *J. pseudosabina* Fisch. et Mey. 2. *J. wallichiana* Hook.f. They regard *J. recurva* Buch.-Ham. a complex of at least 4 taxa.

(i) *J. recurva* (Buch.-Ham.) Mehra et Pathania (*Sensu stricto*) a tree, so far found only in E. Nepal (ii) *J. squamata* var. *fargesii* Rehder & Wilson a tree, E. Himalaya. (iii) *J. squamata* var. *wilsonii* Rehder, shrub, W. Sikkim. (iv) The true *J. squamata,* a shrub widely distributed in the W. Himalaya. For (ii) *J. squamata* var. *fargesii* they suggest it is raised to the status of a species *J. fargesii* Komarov. They also include *J. excelsa* (Grecian Juniper) a tree from Greece, Crimea, etc occuring wild in the W. Himalaya. They recognise 9 taxa in the entire Himalaya. They also mention p. 140 "*J. pseudosabina* Fisch. et Mey. is not synonymous with *J. wallichiana* Hook.f. as mentioned in literature, but the two are distinct occuring in the Indian subcontinent . . . .".

The author regards *J. wallichiana* Hook.f. synonymous with *J. pseudosabina* Hook.f. (non Fisch et Mey). The valid name now for the former species is *J. indica* Bartol.*J. recurva* Buch-Ham. is a widely distributed and variable species. Mehra & Jain's painstaking research on *J. recurva* complex would no doubt interest conifer taxonomists to critically examine their findings.

2. **Cupressus** Linn. Sp. Pl. 1002. 1753. **CYPRESS**

Cupressus or the true cypress is a genus with *c* 20 species in the world with two species native to the Himalaya. They are distributed in N. America, Mediterranean region, Sahara, Himalaya and China. Evergreen trees, very rarely shrubs. *Branches* spreading or ascending terminating in *branchlet systems* or sprays of foliage which are pinnately divided and clothed with closely pressed, scale-like *leaves,* fringed with a narrow transparent border (which can be seen only with a lens of good magnifying power), in four ranks on the ultimate divisions; or awl-shaped on the main axes. *Male and female* strobili appear on the same tree on separate branches. Male strobili cylindrical, terminal on the ultimate divisions of branchlets, bearing numerous stamens, each composed of 2-6 globose anther cells. *Female strobili* consisting of peltate scales with ovules in several rows at the base. The cones are more or less spherical in shape, made up of 6-12 peltate scales, closely fitting together at the margins and ripening in the second year. *Seeds* one-many to each scale, leathery narrowly winged. The most commonly used timber is the Monterey Cypress, (*Cupressus macrocarpa*) of Monterey California which has done well in Australia and New Zealand. It is also planted in the Nilgiris. *C. torulosa* is native to the Himalaya and also the less known *C. corneyana.* The exotics commonly planted in our area along with the native species are keyed out below.

Key to the species

Twigs more or less in one plane

Twigs roundish; leaves truncate with appressed tips................ 1. *C. torulosa*

Twigs more or less flattened

Leaves on the twigs appressed or only the tips somewhat free, bright or grey-green.. 2. *C. funebris*

Leaves on the twigs not appressed, blue-green.................. 3. *C. corneyana*

Leaves blue-green branchlets fan-shaped. . . 4. *C. cashmeriana*

Twigs not in one plane... 5. *C. sempervirens*

1. **Cupressus torulosa** D. Don Prodr. Fl. Nep. 55, 1825 Hook.f. FBI, 5: 645. 1888; Brandis Ind. Trees 693, 1906. Raizada and Sahni 142, 1960; Dallimore and Jackson. Handb. Conif. and Ginkgo rev. Harrison 216 1966. **HIMALAYAN CYPRESS** (Pl. 21, Fig. 2)

W. Him. *Leauri, Devidar, Galla, Surai, Raisal*

Description : A large evergreen tree with a pyramidal crown upto 200′ high with a girth of 36′. Branches horizontal or drooping. *Bark* peeling off in long thin strips. Leaves 2.5 mm long or less, triangular, thick with closely pressed tips so that the branches are not rough to the touch. *Monoecious,* male catkins 5-6 mm long often tinged purple, solitary at the tips of the branchlets. *Cones* globose, scale 6-10. Ripe cones 12-17 mm in diameter, clustered, scales woody. *Seed* 6-8 on each scale compressed about 5 mm across, pale brownish.

Distribution/ Specimens collected : Outer ranges in the Himalaya from Chamba (Himachal Pradesh) to Nepal, Bhutan, Kameng District, Arunachal Pradesh, S.E. Tibet. It also occurs in W. Szechuen China. Its altitudinal range is from 1800-2800 m. The trees in Kashmir and Pakistan are apparently cultivated. They mostly occur on limestone rocks. The trees in Tenga Valley, Arunachal Pradesh are of great age and of immense girth and height apparently due to the high rainfall (Height 30 m, g.b.h. 18 m). *K.C. Sahni* 5178, Rupa Kameng Dt. *C. torulosa* seems to be localised in Rupa and one gets the impression that they are planted. These trees appear to be several hundred years old. UTTAR PRADESH, Chakrata 72000′, *B.B. Osmaston* s.n; *K.C. Sahni* 26601. Fletcher (1976) P. 260, records *C. torulosa* near Tangme 7000′ (Gorges of Tsangpo, S.E. Tibet 'some tremendous monarchs of the Bhutan Cypress, *C. torulosa-Ludlow & Sherrif* 12141 at least 200′ high as straight as a rod...... trunk 36′ in girth ......)'. The Arunachal/Bhutan trees are now identified as *C. corneyana.*

From adjacent Tibet a giant cypress new to science *C. gigantea* Cheng et L.K. Fu has recently been discovered forming small pure forests at 3000-3300 m along the Yarlung Zangbo River. It stands 50 m tall, 5.8 m in circumference and is estimated to be over 2300 years of age. (Jingwei 1982). As so far I have not seen herbarium specimens it is not possible to say how it differs from *C. torulosa.* From the photograph of a full tree they appear to be allied, perhaps more to *C. corneyana.*

Fl. : Male strobili appear in Sept.-Oct.; ripening and pollen shedding Jan-Feb. The trees become conspicuous with golden yellow covering of male strobili. *Fr. :* ♀ cones ripen in the second year. The cones open and shed seeds from Aug.-Dec.

Uses : It is one of the most durable of coniferous woods in India. It is preferable to deodar for internal work such as window frames, ceilings, panel of doors. It is equal to deodar for railway sleepers.

2. **Cupressus funebris** Endl. Synop. Conif. 58. 1847. Dallimore & Jackson rev. Harrison edn 4. 202, 1966.

FUNERAL CUPRESS WEEPING CYPRESS
(Pl. 21, Fig. 1)

A medium sized tree 15-21 m tall, 2.5 m in girth with wide spreading curved and very pendulous branches and branchlets almost making an angle of 90° with the ground floor. It is a native of China and planted in hill stations in the Himalaya, often in cemetries. *Branchlet systems* alternate, flattened, bipinnate, the ultimate division compressed. *Leaves* greyish-green, of 2 kinds, lateral pairs folded face to face; facial pair flattened. *Cones* globose up to 1.2 cm in diameter, on slender, often curved stalks, ripening in the second year, dark brown; scales 8, each with a minute ovate process. *Seeds* 3-5 on each scale, 0.5 cm long, shining reddish-brown with scattered resin tubercles and narrow wings.

3. **Cupressus corneyana** Carriere, Traite, Gen Conif 128 (June 1855) Franco in Port. Acta. Biol. Ser B, 9: 192 (1969) Long, D.G. Notes R.B.G. Edin 38; 311-314. 1980. Grierson & Long Fl. Bhutan Vol. I, pt. I, P. 52, 1983.

C. pendula W. Griff. Itin. Notes 137, sub no. 529, 143. sub no. 679 a (1848) non Thunb (1783).

C. funebris auct non Endl. Hook.f. FBI 5: 646, 1888.

Bhutan *Chendey, Tsenden Shing :* Weeping Cypress.

Tree 30 m, with pendulous branches. Scale-like leaves of branchlets ovate, c 1 mm long, those of main shoots 1.5-2 mm long. Male cones oblong, *c* 3 × 1.5 mm. pale brown, Female cones, when mature 1.5-1.8 cm; scale 6-9 mm across, flat or with a blunt peg at centre.

Distribution : Endemic to Central Bhutan, Pho Chu Valley. Moist forests with Hemlock, Fir and Juniper 2550-3000 m. *Fl. Fr. :* May-July. According to Grierson & Long *loc. cit.* records of *C. torulosa* from Bhutan consider Fletcher (1976) p. 260 *C. torulosa* near Tangme (Bhutan) of this species. Also the huge ancient trees in Tenga Valley, Arunachal Pradesh *K.C. Sahni* 5178 (D.D.) should be *C. corneyana* instead of my earlier determination as *C. torulosa.* They are planted as sacred trees around villages and monastries in Bhutan, Sikkim and Arunachal Pradesh.

4. **C. cashmeriana** Royle ex Carriere Traite Gen. Conif. ed 2: 161 (1867) Raizada & Sahni 143, 1960. Grierson & Long Fl. Bhutan Vol. I. pt. 1, p. 52, 1983.

Similar to *C. corneyana* but smaller **pyramidal-shaped, branches ascending**, branchlets pendulous, fan-shaped, foliage with a bluish tinge. *Cones* greenish yellow with a glaucous tinge when young, dark brown when mature c 1.2 cm in diameter, globose, scales 10, with a central depression and an acute triangular reflaxed process. Seeds *c.* 10 to each scale winged. A very graceful tree.

Distribution : Not recorded in the wild since Royles collections who did not mention its locality in Kashmir. Intensive field work is necessary in Kashmir, particularly, Ladakh, Gurez, etc. cultivated as an ornamental in Dehra Dun, Kalimpong etc.

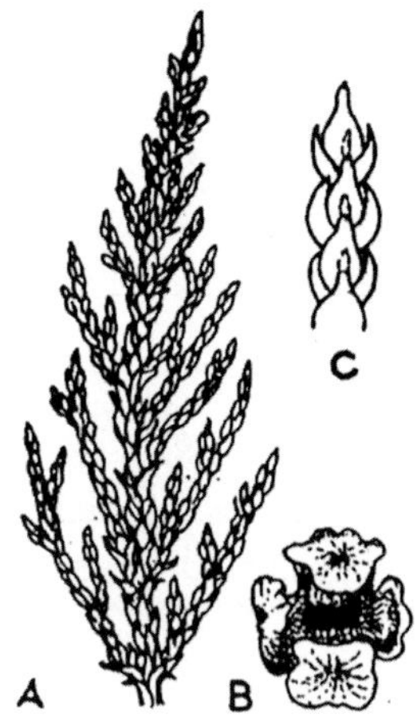

Cupressus corneyana x 1 c x v. B mature female cone. After *K.C. Sahni* 5178.

Uses : It is a very graceful tree worthy of introduction in parks and gardens and as an avenue tree. Fast growing. Dallimore & Jackson rev. Harrison 1966, P. 217 call it *C. torulosa* var. *corneyana* (Knight & Perry) Carriere, but they do not mention its area of distribution.

Fl. & Fr. : Cold season - Fruit not fertile in Dehra Dun.

5. **Cupressus sempervirens** Linn. Sp. Pl. 1002, 1753. **MEDITERRANEAN CYPRESS**
(Pl. 22, Fig. 2)

Etymology : The specific name marks the persistence of the foliage, giving the crown its dark-green colour.

Habitat : Grows wild in regions bordering the eastern Mediterranean. It is widely planted as an ornamental at Agra in the Taj Mahal Garden, Srinagar, Dehra Dun, Simla and in the subcontinent including Burma. Burm. *Kyaukpan.*

Description : A tree 20-30 m/65-100′ with columnar shaped crown. A very slow growing long-lived tree, an age of 2000-3000 being on record. It has an erect trunk and grey-brown *bark,* fibrous and furrowed longitudinally. Branches ascending not whorled, branchlets deep green. *Leaves* with closely adpressed tips, so that the branchlets are not rough to the touch. Male strobili yellow, *c* 2.5 mm long, stamens in c 10 pairs. *Ripe ♀ cones* solitary or few together, c 3-4 cm in diameter; shining brown or greyish, subglobose or ovoid; scales 8-14, rising to a point in the centre or flattened with a thin ridge like process. *Seeds* 8-20 on each scale, c 0.42 cm long, winged, without resin tubercles. The cones take 2 years to mature. It has 2 typical varieties, the spreading form var. *horizontalis* and the pyramidal or columnar form var *sempervirens.* In addition the following species have also been introduced *C. macrocarpa* (Monterey Cypress) of California at Ootacamund. *C. lustanica* Miller (Mexican Cypress) is cultivated in Dehra Dun, Shimla and elsewhere and also makes a good hedge as it is not browsed by animals.

3. **Callitris rhomboidea** R. Br. ex L.C. Richard **CYPRESS PINE**
(Pl. 22, Fig. 1)

C. cupressiformis F. Mueller; *Frenela rhomboidea* Endlicher.

This tree is from the coastal areas of E. Australia and was introduced in the Nilgiris in 1885 on a plantation scale by Gamble around Ootacamund. 9-15 m tall with feathery cypress-like foliage. Cones globular 0.8-1.2 cm. It gives an excellent firewood of the fragrant resinous kind.

4. **Thuja** Linn. *Biota* Endlicher

ARBOR VITAE
(Genus Thuja)

Evergreen trees and shrubs from China, Japan, Taiwan and N. America, all generally known as 'arbor vitae'. *T. occidentalis* L. is the American, *T. plicata* D. Don a tree up to 200′ of W.N. America and *T. orientalis* L. the Chinese arbor vitae. *Branchlets* systems flattened and erect, a field character which distinguishes it from *Cupressus* which are generally pendulous. *Leaves* decussate, scale-like. *Monoecious.* male catkins ovoid, with 3-6 pairs of stamens. *Female* cones ovoid or oblong composed of 8-12 scales which are thickened upwards the 2 or 3 middle pairs fertile. *Seeds* winged or wingless.

In Linn. Syst. ed 1. 1735, *Thuja* is spelt as *Thuya.* According to Art 83 of the International Rules of Botanical Nomenclature, 1952 edition. "If Linnaeus subsequently to 1753-54 consistently adopted one of the spellings that spelling is accepted, e.g., *Thuja* (not *Thuya*)".

Thuja orientalis Linn. Sp. Pl. 1002. 1753, ed. 2 2: 1422, 1763.

CHINESE ARBORVITAE. PEACOCK FEATHERS
(Pl. 26, Fig. 2)

Biota orientalis Endlicher.

Platycladus stricta Spach

Mor Pankhi- Hindi.

Etymology : From the Greek bios, life. Also called tree of life. *Biota* was the other name of the genus.

Description : A bush or a small tree up to 9-10 m. *Branches* and shoots erect (a character which separtes it from Cypress). *Bark* thin red brown. *Leaves* persistent-scale-like and opposite 1.5 mm without glands. *Strobiles* ovoid, fleshy, bear downcurved hooks which protect the 1-2 wingless seeds.

Distribution : N.W. China widely cultivated in India as an ornamental in gardens.

5. **Calocedrus** Kurz in Journ Bot. 11: 196 June, 1873. **NORTHERN INCENSE CEDARS** (Genus) Calocedrus

Heyderia K. Koch, Dendr. 2(2): 179 Nov. 1873 Li in Journ. Arn. Arb. 34(1): 22. 1953.

Libocedrus sub. gen. Heyderia Pilger in Engl. u Prantl. Pflanzenfam. ed. 2. 13: 389. 1926.

Libocedrus sensu Bentham and Hooker pro parte.

The Northern Incense Cedars (CALOCEDRUS) as now understood are distributed in the northern hemisphere. The allied genus *Libocedrus* as now understood is distributed in the Southern hemisphere and are called Southern Incense Cedars. They are (native to New Zealand and New Caledonia) united with *Libocedrus* by some authors. According to Dallimore & Jackson (1966), p. 132 'probably more closely allied to *Thuja* and *Thujopsis. Calocedrus* comprises three species from the Pacific coast of N. America, China, Taiwan and Burma. Only one species occurs in our area in Burma. Evergreen trees. Ultimate *Branchlets* flattened, distichously divided. *Leaves* scale-like, decussate, strongly flattened, approximately equal in length, closely appressed except for the pointed tip, the lateral pairs overlapping the long decurrent facial pairs laterally. *Male* and *Female strobili* solitary, terminal usually on different branches of the same tree, rarely on separate trees. Male *strobili* oblong or ovoid, with 6-16 decussate, sub-peltate, broadly ovate pollen sacs. Female strobili oblong-elliptic, with 3 pairs of erect imbricate, woody, persistent scales, only the middle pair being fertile, each with 2 ovules. Mature *cones* woody, ripening the first year. *Seed* flattened, 2 to each fertile scale, with unequal wings, one about as long as the scale, the other about half as long.

According to Bayard Hora (p. 95 1981) 'the 3 species were formerly included under *Libocedrus* (sensu lato), one character separating them out being that the facial and lateral scale leaves are nearly equal in length'.

Calocedrus differs from *Thuja* in its fewer cone-scales and unequally winged seeds.

Calocedrus macrolepis Kurz in Trimen Journ. Bot. 11: 196 t. 133.f. 3. 1873; Wils. in Journ. Arn. Arb. 7: 63. 1926; Florin in Taxon 5(8): 192. 1956; Dallimore & Jackson Handb. Conif. & Ginkgo rev. Harrison 135, 1966. (Pl. 26, Fig. 1)

Libocedrus macrolepis (Kurz) Benth. in Benth. ex Hook.f. Gen. Pl. 3:426. 1880. Hook.f. FBI: 5, 645, 1890. *Thuja macrolepis* (Kurz) Voss in Mitt. Deutsch. Dendr. Ges. 16: 88: 1907.

Heyderia macrolepis (Kurz) Li, in Li. Journ Arn. Arb. 34 (1): 23. 1953.

Burm. *Tinyuhmwe*

Typus : *C. macrolepis* Kurz.

Description : A tree up to 35m/115′ high, with a broadly conical head. *Bark* whitish scaly. *Leaves* on the main shoots up to 1.2 cm, ending in a small spine. those of lateral branchlets 0.6-0.8 cm long, glacous beneath; the keel of each lateral leaf straight or curving outwards near the sharply pointed apex (giving the shoot a jointed appearance). *Facial leaves* obtrullate or obtriangular, midrib raised with a central groove conspicuous above, midrib not prominent beneath, terminating abruptly in a sharp point. *Cones* ellipsoid 0.6-1.2 cm long, on slender, 4-sided shoots up to 3 cm long. *Seeds* usually one to each fertile scale.

In general appearance it resembles *Thuja orientalis* L.

Distribution : A rare taxon native of S. Yunnan, Hainan and Mong Kaing in S. Shan States (Burma), occurring in ravines near water courses at altitudes of 1200-1500 m. Its discovery from Mong Kaing is an interesting record in Burmese Flora.

Specimens examined: Burma Mong Kaing 7-8000′, S. Shan States, Aug 1899, *H.N. Thompson* s.n. Fruiting : Yunnan S. Shan States Frontier, Kengtung, Tree 8′ grith. Dy. C.F. Reg No. 23906. Det Kew 1921.

Fl. : April-May. *Fr. :* Aug.-Sept. (Herbarium sheets in DD). This taxon was till recently considered to be found in Taiwan also. Florin (1956) calls the Taiwan tree as *Calocedrus formosana* (Florin) Florin syn *Libocedrus formosana* Florin. This is very close to *C. macrolepis* in form but its leaves are 2 mm long, with two stomatal lines on the lateral leaves, whereas there are four in *C. macrolepis.*

Uses : The wood is useful for many purposes. In China, buried trees are dug up and their wood is in demand for making coffins. (Dallimore & Jackson rev. Harrison loc. cit.). It is creditable for Sir Joseph Hooker to predict its occurence in Burma. In FBI V: p. 385, 1886 he says the Yunnan *Calocedrus macrolepis* Kurz, may be expected to be found in the mountains of N.W. Burma.

Till recently three genera, *Gnetum, Ephedra* and *Welwitschia* made up the order Gnetales. As these three genera possess naked ovules, they are undoubtedly gymnosperms. The discovery of a new taxon *Sarcopus aberrans* from Indo China reported by Gagnepain (1946) added a fourth genus to the *Gnetales.* It is characterized by the presence of a single naked ovule. Gagnepain separated it from other members by its bisexual flowers whose female element is represented by a naked ovule and treated this monotypic genus under a new family *Sarcopodaceae* which is considered to be the most advanced of the order.

Only *Ephedra* and *Gnetum* are represented in our area. The *Gnetales* differ from other gymnosperms by their compound staminate strobili, opposite or whorled leaves (also found in *Cupressaceae*), vessels in the secondary wood and by absence of resin canals. Only the naked ovule character keeps this order in the gymnosperms and only *Ephedra* has a gymnospermous life history. They possess many characters common to the angiosperms and have been interpreted by some botanists to be a partial connecting link between the two. However no *Gnetales* have been found as far back as the cretaceous, when angiosperms were abundant and there are important morphological considerations that lead some botanists to believe that the *Gnetales* are themselves an isolated assemblage whose genera are remotely related to one another and far less so to either the gymnosperms or angiosperms.

Gnetum, Ephedra and *Welwitschia* were originally placed in a single family *Gnetaceae.* This heterogenous grouping drew the attention of leading botanists of the time. Bertrand (1874) remarked that the three genera "forment des groups absolument separes". By the turn of the 19th century the order was split up into three families with one genus in each. Markgraf (1926), monographer of this group considered that the order Gnetales is comprised of three families *Gnetaceae, Welwitschiaceae* and *Ephedraceae.* This view was supported by Sporne (1965). This treatment is also followed in the present work.

Arnold (1948) and Lam (1951) treated them under Chlamydospermophyta and Chlamydospermae respectively. Florin (1934 a) and Eames (1952) have established three orders, *Ephedrales, Welwitschiales* and *Gnetales* with a monogeneric family in each. Pant (1957) placed *Gnetum* and *Welwitschia* under *Chlamydospermophyta* and treated *Ephedra* under *Coniferophyta.* Foster and Gifford (1959) placed them under a single order *Gnetales.* According to Thierot (1980) p. 523 'The characteristics that unify the GNETACEAE are (1) The presence of vessels in the wood (2) the compound strobili, both pollen and seed (3) the presence of a perianth superficially corolla or calyx-like, associated with pollen producing and seed producing structures; and (4) the presence of two integuments, the inner one produced into a long tubular microphyle . . . . ."

Origin and affinity :

The origin and affinity of the *Gnetales* are not certain and have evoked considerable interest for more than 50 years (Dutta 1973). According to Pearson (1929), the relationships of the three genera of the *Gnetales* are perhaps as obscure now as they have been at any time within the last decade. The fact that the 3 probably have great antiquity is apparent from their present distribution pattern and morphology. (Khoshoo and Ahuja, 1963). The *Gnetales* is an admixture of gymnosperms and angiosperms traits. The naked ovule reminds one of the gymnosperms. They bear vessels in the secondary wood like the angiosperms also the compound strobili are borne in flower-like groups surrounded by sterile bracts which resemble the sepals and petals of angiosperms. Maheshwari and Vasil (1974) consider them as 'notable among the gymnosperms on account of the presence of certain angiospermic features. They have often been called the lure and despair of the morphologist. They are alluring because they seem to give some hope in the solution of the morphologist's great problem, the origin of the angiosperms. They are a despair in that, so far, no one has been able either to establish convincingly or to disprove their angiospermic connection'. Foster and Gifford (loc cit) regard them as 'extremely specialised end-points in evolution, because of certain similarities in their organography, anatomy and reproduction'. The importance of the *Gnetales* to the solution of problems of angiosperm origin have been emphasized by Markgraf (1930), who held the angiosperms to be a taxon derived from gymnosperms, and one that branched off from ancestors of modern gymnosperms so far back in geological history and bear no particular relationship to present-day conifers. Fagerlind (1947) reinterpreted the morphology of the *Gnetum* flower, demonstrated it be homologous with those of *Ephedra* and *Welwitschia,* and considered the 3 taxa to have had a common ancestor which was derived from th same stock as the single ancestor postulated to have given rise to the Pro-angiosperms from which all recent angiosperms evolved polyphyletically.

Classification : The scheme of classification for Gnetales in the present account is based substantially as that proposed by Melchior and Werdermann (1954) in Englers Syllabus der Pflanzenfamilien.

A. CYCADOPSIDA	B. CONIFEROPSIDA	C. GNETOPSIDA.
1. Pteridospermales	1. Cordaitales	1. Gnetales
2. Bennettiales	2. Coniferales	
3. Pentoxylales	3. Taxales	
4. Cycadales	4. Ginkgoales	

Gnetales :

Woody plants, trees (only one in our area), shrubs or liana. Leaves opposite, whorled, simple, broadly elliptic, with netted venation, in aspect angiosperm like (*Gnetaceae*) or strap shaped (*Welwitschiaceae*), or reduced to minute scales (*Ephedraceae*). Secondary wood with vessels. 'Flowers' unisexual and normally dioecious (except some monoecious species of *Gnetum*). 'Flowers' organized into a compound strobili or inflorescence. Female flowers with a single erect ovule, the nucellus of which is surrounded by two or three envelops, the micropyle projecting as a long tube. Male flowers with a 'perianth' and antherophores with one to eight synangia. Fertilization by means of a pollen tube with two male nuclei. Embryo with two cotyledons. (Sporne 1965, modified).

Gnetaceae	:	*Gnetum*
Welwitschiaceae	:	*Welwitschia* (Restricted to the coastal belt of S.W. Africa)
Ephedraceae	:	*Ephedra*
Sarcopodaceae	:	*Sarcopus* Indo-China

The two families in our area are easily separated by the following key.

Shrubs/liana (rarely a tree) with broad evergreen opposite leaves - *Gnetaceae*

Shrubs/climber with opposite connate leaves reduced to scales - *Ephedraceae*

Family 10. GNETACEAE

Lindl. Bot. Reg. 20. 1686. 1843. **GNETUM FAMILY**

Most are climbing shrubs, a few erect shrubs, one a small tree in our area. *Leaves* decussate, exstipulate, simple, evergreen leathery. *Flowers* dioecious, in spikes, which are frequently grouped into more complex inflorescences. The spike bears decussate bracts in whose axils are condensed partial inforescences of a large number of flowers as in *Labiateae* *c* 3-8 in the female but more up to 40 in the Male. These flowers form whorls round the stem, and are inter-mingled with numerous hair structures. At the top of each nodal group of the Male inflorescence in most is a single ring of the female flowers usually with only one integument and infertile sometimes with 2 or even 3 integuments and fertile. The male has a tubular 2 leafed perianth, from the top of which the axis projects; at the tip of the axis, right and left, are two sessile 1 locular anthers. The female has a tubular perianth like that of *Ephedra,* surrounding a single orthotropus erect *ovule* with two integuments; the inner of these projects at the apex of the flower. But there is much difference of opinion as to the morphology of these envelopes. After fertilization the perianth becomes fleshy, the outer integument woody, forming a drupe-like *fruit.* There is only one genus *Gnetum* in this family.

Gnetum L.

Distribution : There are up to forty species of Gnetum in the world most of which are lianes, although a few, such as *G. gnemon,* are trees, and some are shrubs. The genus in confined to the tropical, monsoon and equatorial regions of the world, namely Equatorial S. America, W. Trop. Africa, India including Andamans and Nicobars, Bangladesh, Burma, Malaya Peninsula, Indonesia, Philippines, New Guinea, Thailand, Indo-China, and China. It is absent from Central America, Sri Lanka and Australia. Some of the Asiatic species show a wide distribution and extend as far as the Solomons. Malaya, Borneo and the Indo-Burmese region have the maximum number of species, while large continents like Africa and S. America are comparatively poor. The largest concentration of endemics is in S. America and the highest number of indigenous species are in Malaya and Borneo. None of the Asiatic species reach Africa or America. (Bharadwaja 1957, Sporne 1965).

According to Airy Shaw (1973) in rev. Willis. Dict. Fl. Plants & Ferns edn. 8, p. 497 *Gnetum* has 30 species in the world. He adds Fiji to the list above.

Habitat : All species inhabit the tropical rain forest below an altitude of 1500 m except *G. microcarpum* which has been reported from Malaya at *c* 2000 m. Species with large leaves and fruits are found in the moist air of mountain gorges and in cloud forests while those with small leaves occur both in the rain-forest and in drier localities (Markgraf 1954).

G. gnemon is cultivated in Malaysia for its edible seeds. Of the many interesting features shown by *Gnetum,* the most remarkable is the appearance of the leaves, which are indistinguishable from those of angiosperms in having a broad lamina and reticulate venation.

Inflorescences : All the species are dioecious and the inflorescence is generally a panicle, which is either solitary and axillary or fasicled. Terminal cones are also met with sometimes. In some species the female, more rarely the male, panicles are cauliferous.

The cone consists of a stout axis which bears at its base two opposite and connate bracts in the axils of which other inflorescences, buds or flowers may be formed. Higher up whorls of circular bracts are present and arranged one above the other to form cupules or collars. Initially the young cones have a compact appearance when the internodes are short. Later the internodes elongate and the collars are separated from one another. The number and shape of the collars and the length of the internode separating them are characters of diagnostic value. In *G. ula* the cone terminates in a bare segment. In *G. gnemon* the upper two or 3 collars may be very much reduced and sterile. (Maheshwari and Vasil 1974).

Male cone : The male flowers are arranged in definite rings usually three to six in number, above each collar. Just above the male flower there is a single ring of imperfect female flowers or abortive ovules. Ovules are rarely present in the male cone. The male flower, which is enclosed in a sheath-like perianth consists of a stalk bearing two unilocular anthers. When the anthers are mature and ready for dehiscence, the stalk elongates and pushes them through the slit in the perianth beyond the collars of the cone.

In *G. ula* and *G. gnemon* the number of anthers has been found to vary from two to four (Vasil 1959, Madhulata 1960).

Female cone : The female cones are very similar to the male except that a single ring of four to ten female flowers or ovules is seen above each collar and very few ovules develop into mature seeds. Each ovule or female flower consists of the nucellus surrounded by three envelopes. The inner envelope elongates to form the micropylar tube or style which extends beyond the middle envelope. In the mature seed the middle envelope is stony and the outermost is fleshy. In *G. gnemon* and *G. ula* some abnormal female cones have been seen in which the collars are arranged spirally (Maheshwari and Vasil 1974, Sporne 1965).

In both male and female cones, the flowers are interspersed with uniseriate hairs.

The first comprehensive taxonomic account of Gnetum for the subcontinent of India including Burma is by Sir Joseph Hooker published in 1890 in the fifth volume of the Flora of British India. Earlier the species were imperfectly known and there are references to the genus by Rheede (1688), Roxburgh (1832), Brandis (1874) and Kurz (1877). The first modern investigation is by Markgraf (1930) whose researches have modified much of our knowledge regarding the taxonomy of the genus. Based on these findings Bharadwaja (1957) published an account of the Indo-Burmese species. During the last 3 decades the botanists of the Forest Research Institute, Dehra Dun and its regional centres and the Botanical Survey of India have discovered new localities and new range extensions for the genus. The specimens examined are mostly from the Herbarium of the FRI, Dehra Dun, its regional centres and specimens in the Central National Herbarium Calcutta which were loaned to me. The present study includes recent collectors and localities subsequent to Bharadwaja's (1957) publication.

Brief Generic/detailed species account :

Gnetum Linn. Mant ed I (1767); Hook.f. FBI, 5 1890 Markgraf in Engler & Prantl Die Nat. Pflanzenfam 2 Auf. 13: 440. 1926 & Bull. Jard. Bot. Btzg 111, 10 (1930) 407. Bharadwaja, R.C. *J. Indian bot. Soc.* **36**: 408-420, 1957; Maheshwari & Vasil, Botanical Monograph No. I, Gnetum, CSIR, New Delhi, 1974.

Shrubs or mostly lianes, rarely trees. Branches with swollen nodes. *Leaves* petiolate decussate, simple with reticulate venation. *Inflorescence* a spike or a panicle of spikes, axillary, or terminal; axis stout bearing cupular bracts, each subtending one or more rings of flowers, terminal segment usually with a whorl of sterile female flowers. *Flowers* usually dioecious rarely monoecious surrounded at the base with numerous hairs. Male Flowers : perianth claw-shaped, bifid; microsporophyll one, adnate to the base of the perianth and exerted; microsporangia 2, dehiscing transversely. Flowers sessile or subsessile; perianth tubular and thick, ovule ovoid or globose; integuments two in fertile flowers, one in imperfect flowers, inner integument produced into a slender exserted micropylar tube with a fimbriate mouth. *Fruit* drupaceous, sessile or peduncled with a fleshy or fibrous epicarp, hard mesocarp and papery endocarp.

Type : *Gnetum gnemon* Linn.

Distribution : About 40 species of which 7 are in northern S. America, 2 in western tropical Africa, the remaining in tropical Asia from the western ghats through Malaysia to Fiji, none in Taiwan and new Caledonia.

Ecology : All species inhabit the tropical rain forests below 1500 m, also found in Burma. Species with large leaves and fruits are found in the moist mountain gorges and cloud forests while those with small leaves occur both in the rain forests and in drier localities (Markgraf 1954).

The pink fruits of *Gnetum* are dispersed by birds and seeds and are sometimes found in the excreta of civet-cats (Markgraf 1954). The seeds are sought by mammals and birds as a food source (Thierot 1980). Eight species and four varieties are known from the subcontinent including Burma. These are described giving their keys both for male and female plants. Bharadwaja (1957) has given workable keys which are also followed in the present work with minor changes.

Markgraf (1930) divided the genus into two sections and five subsections of which the section Cylindrostachys is the largest and most of the species of our subcontinent including Burma belong to this subsection.

Key to the Sections

1. Male inflorescence with long internodes, axis visible between the bracts (except in *G. gnemon* Linn. var. *griffithii* Mgf.) .... 1. Gnemonomorph

1′ Male inflorescences with short internodes, axis not visible between the bracts .... 2. Cylindrostachy

Section I. *Gnemonomorphi*

This section contains 10 species and 8 varieties confined to Africa, Asia and America. The section is divided into three subsections viz. 1. *Eugnemones,* 2. *Micrognemones* and 3. *Araeognemones,* of which the first is represented in the Indo-Burmese area by *Gnetum gnemon.*

Subsection : *Eugnemones*

Leaves thin. Trees and shrubs. Dried leaves and inflorescence yellow. Fruits finely velvety.

1. **Gnetum gnemon** Linne, Mant. 1 (1767) 125: Roxb; Fl. Ind. 518, 1832; Kurz For. Fl. Brit. Burma 2: 497, 1877; Hook.f. FBI 5: 461, 1890; Brandis Ind. Trees, 687, 1906; Markgraf, Bull. Jard. Bot. Btzg 111, 10 (1930) 436; Bharadwaja *J. Indian bot. Soc.* 36: 412, 1957; Maheshwari & Vasil, Bot. Monograph I, Gnetum, CSIR, New Delhi, 1974. (Pl. 27, Fig. 1)
   Burm. *Tanyinwe, Hinbyin.*

Small evergreen trees or shrubs, crown narrow conical with drooping branches; trunk grey, marked with conspicuous or faint rings. *Leaves,* thin, yellow when dry, narrow at both ends, varying in shape and size, broadly elliptic, acuminate, 7.5-20 × 2.5-10 cm; petiole 6-18 mm. ♂ *Inflorescences* solitary, axillary, simple or once branched, yellowish 3-6 cm long, collars 3 mm long; sporangia 2, globose, yellow; *imperfect* ♀ *flowers* 10-15, globose 2-2.5 mm thick, apiculate or beaked; perianth fleshy, integument thick and fleshy. ♀ *Inflorescence* similar to the male flowers 5-8 at each node, globose, tipped or beaked, 3-4 mm long, inner tube exserted by 1 mm. *Fruit* oblong, apiculate 1-2.5 cm long; glossy, orange-coloured or red when ripe, containing a single large starchy seed.

Localities : ARUNACHAL PRADESH, Perila Kameng Dist. 16081; ASSAM Potfal, Kacharigram Reserve Forest, *Panigrahi* 5450, Garampani & Barapathar *Panigrahi* 9427; MEGHALAYA, Jowai-Badarpur Road 10984; TRIPURA Kawnpuri *D.B. Deb* 27352; NICOBARS, Kadoo village, Katchal Island *Chakraborty* 5267, Passa, Car Nicobar *N.G. Nair* 3516, Sawai, Car Nicobar *N.G. Nair* 2617 (30.5.1975); Gt Nicobar Island, Campbell Bay *Thothathri & Bannerjee* 11355, Galathea River, *Thothathri & Bannerjee* 11504. Tree 15 m tall. *See* Thothathri *et al* 1973.

From Thothathri specimens cited above with field notes, it is apparent that the tree form in our area is mostly in the Gt. Nicobar Island. The photograph of the tree from Indonesia forms the frontispiece in Maheshwari & Vasil's Monograph (1974). This tree should be similar to the Gt. Nicobar tree as this island is adjacent to Indonesia. The shrub type is characterized by the following features. Old collections and localities are cited below.

Shrub 0.6-3 m. Collars remote only the uppermost ones contigous, axis unbranched. Flower shortly tipped; fruit small, 1 cm long, characterized by sessile seeds borne on a thick axis.

Localities : ASSAM, South Lushai Hills (now called Mizoram) 1-1500 m. *Gage* 269 (Herb DD); Naga Hills, Golaghat *King* (Herb CAL). BURMA : Katschin (Kachin) Hills, *Shaikh Mokim* (Herb CAL). *Proudlock* 65 (Herb DD).

Distribution : Malaya, Anamba Isls and N.W. Borneo.

var. **griffithii** (Parl.) Mgf. l.c. 442. *G. griffithii Parl* in in DC. Prodr. 16, 2: 349, 1868. (Pl. 28, Fig. 1)

Shrub up to 2 m. *Leaves* oblong usually with parallel margins. *Inflorescence* contracted, the axis is hardly visible between the bracts. Imperfect and perfect *flowers* gradually narrowed to an acute beak. *Fruit* small, globose, shortly apiculate.

Locality : ASSAM, Sibsagar, *U.N. Kanjilal* 3833 Herb (CAL); Naga Hills, Kungba 1500 m, *Meebold* 7425. vide Markgraf.

Ecology : Ascends into the mountain rain forests in Assam, Naga Hills and Mizoram up to 1500 m.

BURMA : Tavoy, Tenasserin River, *Parker* 2409. (DD), S. Tenasserin, Nagawun. Chaung forest, *Parkinson* 1630. (Herb DD), Mergui *Meebold* 14302 Herb CAL.

Uses : The seeds of *G. gnemon* are eaten after roasting or cooking. The orange red pulp is removed and the seed mashed into cakes, dried and fried in oil into a cake or biscuit. The bark yields a strong fibrs durable in sea water and therefore valued for fishing lines and nets. (Wealth of India 4: 157, 1956).

G. gnemon L. var. *gnemon* is cultivated in Malaya and Java for its fruits which are edible.

Section 2. *Cylindrostachys*

The inflorescences without visible internodes between the collars. Dried leaves never yellowing. Fruits never valvety. Lianes, 17 species and 5 varieties distributed in India, Malaya, Indo-China, E. Indies and New Guinea. Divided into two subsections. *Stipitati* and *Sessiles* Both represented in our area.

KEY TO SUBSECTIONS

1. "Fruits" stalked .... 1. *Stipitati*

1′. "Fruits" sessile .... 2. *Sessiles*

1. *Stipitati*

Bracts of the male spikes flat. Their margins bent outward and exposing the flowers (except *G. contractum* Mgf and *G. latifolium* Bl. var. *macropodum* Mgf. Fruits stalked.

Species 7 and varieties 6. Five species within the area of this work.

Key to the Species
(From Bharadwaja 1957, modified)
(Male plants)

1. Bracts of the male spikes flat, with margins bent outwards and exposing the flowers.

 2. Male inflorescence usually simple.

  3. Leaves brown on drying; lateral veins distinctly jointed; tertiaries distinctly and faintly reticulate on the lower surface, lamina 18 × 8 cm .... 2. *G. oblongum*

  3′. Leaves black on drying; lateral veins indistinctly jointed before the margins, tertiaries reticulate, thin, distinct on the lower surface; lamina 8-15 × 4 - 9.5 cm. .... 3. *G. ula*

 2′. Male inflorescence usually branched .... 4. *G. montanum*

1′. Bracts of the male spike cylindrical, margins not bent outwards, flowers included.

 4. Fertile parts of the male spikes upto 1.5 cm long; inflorescence short. 5. *G. contractum*

 4′. Fertile parts of the male spike up to 4 cm; thick inflorescence lax up to 12 cm long. 6. *G. latifolium*

(Female Plants)

1. Female inflorescence simple.
   2. Fruit smooth on drying, not pruinose, 3 × 1.2-1.5 cm .... 2. *G. oblongum*
   2′. Fruit longitudinally rugose on drying, pruinose 3-4 × 1.5-2 cm .... 3. *G. ula*

1′. Female inflorescence branched.
   3. Fruit longer than its stalks.
      4. Leaves 30 × 12 cm; inflorescence branched twice or thrice, fruit ellipsoid .... 4. *G. montanum*
      4′. Leaves 10 × 5 cm; inflorescence branched only once, fruit oblong-ovate acuminate .... 5. *G. contractum*
   3′. Fruit shorter than its stalk. .... 6. *G. latifolium*

2. **Gnetum oblongum** Mgf. l.c. 471; *G. funiculare* Kurz, For. Fl. Brit. Burma 2: 496. 1877; Hk.f. FBI 5: 643, 1890 (non Blume). Bharadwaja *J. Indian bot. Soc.* 36: 408-420. 1957. (Pl. 29, Fig. 3)

A scandent shrub. Branches slender. *Leaves* coriaceous, brown on drying, lamina oblong-elliptic, cuspidate or apiculate, up to 18 × 8 cm; secondary veins distant, curved, tertiaries distinctly and finely reticulate beneath, petiole grooved 2 cm long. *Male inflorescence* simple, spikes 3 cm × 4 mm broad; flowering bracts pateliform. *Male flowers* many, perianth narrowly obconic 1.5 mm × 0.5 mm; sporophyll 3 mm long, sporangia 2, oblong; *imperfect female flowers* few, conical 1 mm × 0.5 mm; perianth fleshy. *Female inflorescence* simple up to 8 cm; bracts patelliform. ♀ Flowers up to 6 in each node, oblong surrounded by dense rusty hair, perianth 3 × 1.5 mm. *Fruit* oblong, obtuse, apiculate, red, 3 × 1.5 cm, epicarp fleshy and fibrous, mesocarp thin, woody, endocarp chartaceous, seed oblong 2 cm long.

Localities : BANGLADESH (Chittagong, Demagiri) *Lister* 236 (Herb CAL). BURMA: Tonkyeghat. *Kurz* 494 (Herb CAL). Tavoy, Wazum Chaung, *Parker* 2301 (vide Markgraf) Mergui, Lemji, *Parker* 2669 (vide Markgraf).

Distribution : Endemic in the Bangladesh-Burmese region.

On this species Markgraf (1930) comments "very probably this is the species called *G. scandens* by Roxburgh, for his *G. scandens* came from

Chittagong, where apparently only this species occurs, and where it is frequent. But this is not certain, because no original specimen can be traced, and Roxburgh's description has a printing error, at the most important place where he speaks of the fruits (one line is left out and a wrong line repeated). But since he knew only one liana-type *Gnetum*, he erroneously included *Ula* Rheede and *Gnemon funiculare* Rumph. Since then the various plants have been called *G. scandens* Roxb. Therefore I introduce a new name for the species of Chittagong, to be on the safe side".

3. **Gnetum ula** Brongn in Duperrey, Voy. Monde 7: 12. 1829. Mgf. l.c. 469; c. Fisch Fl. Madras 3: 1885 (1304) 1936; Bharadwaja. *J. Indian bot. Soc.* 36: 408-420, 1957; Sant Fl. Khandala ed. 3, 261. 1967; Mathew & Rani Fl. Tamilnadu Carnatic pt. 2. 1916, 1983. *G. funiculare* Smith ex Wight, Icon. pl. Ind. Orient. 1955. 1853, non Brongn. 1829. *G. scandens* auct. non Roxb. : Hook.f. FBI 5: 642. 1888. (Pl. 30, Fig. 4) (Pl. 31, Fig. 5)

TAM-*Ana pendu;* MAL-*Odal, ula,* KAN-*Kodkanballi;* ORIYA -*Lolari,* BOMBAY - *Kumbal, umbli.*

A large dioecious liane 15(25) m. *Bark* thick scaly. Branches with swollen nodes. *Leaves* elliptic-ovate, acute and coriaceous. 8-17 × 4-9 cm; lateral veins 6-8 pairs, curved, tertiaries reticulate, thin, distinct beneath, petioles 1-1.5 cm. Leaves turn black on drying. *Panicles* axillary, arising from stout, warted nodes; strobili arising from the axils of a pair of basal, opposite bracts, successive bracts (collars) decussate, cupular, 3-5 mm apart, compact, elongating with age, ♂ strobilus : 4-5 cm long; ♂ *'Flowers'* 20-30 (40) in a ring, embedded when young, exserted later. Bracteoles c 2.5 mm, top convex or pyramidal. *Stamen* 1 exserted; filament 1 mm; anthers (micro-sporangia) 0.6 mm. ♀ strobilus 5-7 cm long. ♀ *'flowers' c* 8 around node, exserted when mature. *Ovary* (megasporangium) globose, 2 mm across, tipped with a shallow depression. *Fruit* ellipsoid3-4 × 1.5-2 cm, longitudinally rugose on drying. Epicarp fibrose, mesocarp thin woody; endocarp chartaceous. *Seeds* oblong 2 cm long. Stalk 0.3-1 cm. Mostly in hill forests above 600 m.

Leafing, flowering and *fruiting :* New *leaves* in March. Evergreen *Flowers* March-April. *Fruits* April onwards, ripening yellow in November.

Distribution : On the Western ghats in hill forests above 600 m. Nilgiris on forest trees. Lower Burma (Lace, rev Rodger p. 168, 1922).

MAHARASHTRA, Bombay, Yellepur *Bor* (Herb DD), Khandala, *Braganza* (Herb DD), Poona *Woodrow* 16 (CAL). Goa, Chota Dudhsagar *Sahni* 1692, Molem *Sahni* 6088, Karanjol *Bahadur* 2337, Sanguem (S Goa) *Bahadur* 2739. Herb DD). Valpoi. *Kanodia* 96416. Herb Cal.

KARNATAKA. *Saldanha* 16404, Manachina halla, Bedguli *A.S. Rao* 7996. KERALA. *K.N. Subramanian* 2731 (DD), Nilambur *Vaid* (DD), Quilon *Wight* 2757. TAMIL NADU Sirumalai, Madurai dist. *K. Subramanyan* 5795, Nilgiris *Fischer* 2595 (CAL). ANDHRA PRADESH, Modalamma-gudi, Visakhapatnam Dist. *G.V. Subba Rao* 47351 (CAL), Godavari Dist. (vide Markgraf). ORISSA Ganjam, Mahendragiri 1300 m *Fischer & Gage* (Herb CAL). ANDAMANS, S. Andamans *Kirat Ram* (DD).

Uses : The fruits are edible.

4. **Gnetum montanum** Mgf. l.c. 466. Bharadwaja *J. Indian bot. Soc.* 36: 408-420. 1957. Hara et al En. Fl. Pls. Nepal 1: 23, 1978. *G. scandens* (auct. Plur) Hk.f. FBI 5: 642 (1890) pp. *G. edule* Kurz (non Blume) For. Fl. Brit. Burma 2: 495, 1877. (Pl. 32, Fig. 6)

Ass - Mamelet; LUSHAI - *Thanlping-rhui.* KHASI- *Mai-lar-iong-um.*

A large evergreen dioecious climber. Branches smooth, slender, swollen at the nodes. *Bark* peeling off in pieces. *Leaves* dark green, turning black on drying, oblong-ovate 30-12 cm (sometimes smaller); secondary veins distant, tertiaries indistinctly reticulate, petiole 1-1.5 cm long. Male *Inflorescence* branched twice, branches aggregate at the nodes up to 8 cm long, fertile parts of the spike 3 cm long. *Flowers* up to 20; perianth 1 mm, sporophyll 1 mm exserted, sporangia 2, small globose. *Imperfect* ♀ *Flowers* ovoid 1 mm high, integument with cleft apex. ♀ *Inflorescence* branched twice or thrice, bracts 5 mm distant from each other, 4 × 1 mm. ♀ *Flowers* 5-7 in each bract, ovate, perianth 3 mm long 2 mm thick, inner integument produced into a slender, projecting tube with toothed mouth. *Fruit* ellipsoid 1.5 × 1 cm, shining, smooth, shortly stalked; 'epicarp' red, fleshy, thin, fibrous; mesocarp 'woody; endocarp chartaceous. *Seed* oblong.

Distribution : BIHAR Thalakabad *G.V.S. Rao* 23222 1882 (Herb DD); ORISSA, enroute Meghanada Parbat *Panigrahi* 8587; Kotakhal *Deka* 5229. BENGAL - Darjeeling *Gamble* 9814, Jalpaiguri *S.K. Mukerjee* 557, SIKKIM 1500 m without locality *King* 977 ASSAM Patfal *Panigrahi* 5451, Dulang, *Panigrahi* 27710, Chirang *R.S. Rao,* Mowryn Khalm *Dekka* 5259. ARUNACHAL PRADESH 8 km W Tezu, Lohit Dt. *A.S. Rao* 47929, Sissni *Panigrahi* 6168, Khopa, Lohit Dt. *A.S. Rao* 48121, Chowkhan Kheram. Lohit Dt. *J. Joseph* 48356. ANDAMANS Interview Island *Thothathri* 9214. BURMA, Taungnyo Reserve, moist shady ravines *Rogers* 275, Katha Dist. 233 m *Rogers,* Pegu *Kurz,* Tharawaddy *Rogers* (Herb DD), Martaban *Kurz* 2220. Thailand, China & Indo-China. (NEPAL *Williams & Stainton* 8295. E. NEPAL *Stainton* 475, 300-1800 m.

f. megalocarpum Mgf l.c. 458. (Pl. 33, Fig. 7)

Leaves large. Fruit 3 × 1.8 cm, stalk 8 mm long.

Locality : ASSAM (Daphla (Dafla) Hills *Lister* (Herb CAL). The above locality Dafla Hills is Kameng Dist. Arunachal Pradesh, after reorganization of old Assam State.

Distribution : So far endemic in Arunachal Pradesh, Assam.

5. **Gnetum contractum** Mgf l.c. 470; Gamble Fl. Madras 3: 1885, 1934, *G. scandens* Hk.f. FBI 5: 642, 1890 non Roxb. (in part). Bharadwaja l.c. 417. (Pl. 34, Fig. 8)

Scandent shrub, branches cylindrical. Leaves turn black on drying. *Leaves* elliptical, 10 × 5 cm, petiole 1 cm long. Male *Inflorescence* short, once branched, fertile parts usually sessile up to 1.5 cm; bracts cylindrical dense, 2 mm high. *Flowers* many obconic 1 mm long, sporophyll with 2 sporangia. *Imperfect* ♀ *Flowers* up to 10, oblique, oblong, apiculate integument deeply cleft. *'Fruit'* oblong-ovate, long acuminate 3-3.5 × 1.5 cm, smooth, stalk 3-6 × 2 mm; epicarp 1 mm thick, fibrous; mesocarp thin woody; endocarp shining, chartaceous. *Seed* oblong 2.5 cm long.

Not easily distinguishable from *G. ula* in vegetative condition.

Localities : KERALA. Quilon *Wight* 2755 (vide Markgraf) TAMIL NADU. Nilgiris, Coonoor *Perrottet* 308 (vide Markgraf).

Distribution : Endemic to Kerala and the Nilgiris up to 1500 m.

6. **Gnetum latifolium** Bl. in Tijd. Nat. Geschied & Phys 1: 160. 1834; Mfg op. cit. 450; Bharadwaja op. cit. 417.

Large liana. *Leaves* dark green, turning black on drying, leathery, 25 × 12 cm; elliptic-ovate, shortly acuminate, petiole 1 cm, usually grooved. Inflorescence lax, branched, terminal, much branched when cauliflorous, up to 12 cm long, fertile parts of the spike 4 cm long, 4 mm broad, their collars open. Flowers numerous (*c* 50), obconic, 1.5 × 1 mm, sporophyll 3 mm long. *Sterile Flowers* 6-8 in each collar. *Inflorescence* similar up to 15 cm long, their spikes 6 cm long, their collars 3 mm spread. *Flowers* 6-9, acuminate and bent upwards 4 mm long, inner envelope deeply split. *Fruit* pink, ellipsoidal 1.5-2.5 × 1-1.5 cm, distinctly stalked (axis of inflorescence up to 30 cm), outer envelop shining, fleshy, fibrous, 2 mm in diameter, middle one hard, thin, inner one papery. *Seed* broad-oblong.

Key to the Varieties (After Bharadwaja 1957).

1. Leaves black in dried specimens; veins indistinctly reticulate, joined near the margins .... var. *macropodum*

1′. Leaves brown in dried specimens; veins distinctly forming a coarse reticulum on the lower surface .... var. *funiculare*

var. **macropodum** (Kurz) Mfg op cit 462; Hook.f. FBI 5: 643, 1890 (Pl. 35, Fig. 9)

Leaves large, base more or less cordate. *Inflorescence* cauliflorous, up to 30 cm long; male spikes with cylindrical bracts.

Fruit : stalked, stalk equal or longer than the fruit.

Localities ; ANDAMANS. Port Blair *King* (Herb Kew vide Markgraf), Mt. Harriat *Kurz* (Herb. CAL); S. Andamans *Parkinson* 1067 Herb DD. NICOBARS Kamorta *Kurz* (Herb Kew vide Markgraf).

Distribution : As above.

var. **funiculare** (Bl.) Mfg op. cit 463; *G. funiculare* Bl. (Pl. 36, Fig. 10)
Bl. Tigd. Nut. Geschied & Phys 1: 162. 1834,
Bharadwaja op. cit. p. 418. Characters as in Key.

Locality : SOUTH ANDAMANS, Port Monat *King* (Herb CAL).

Distribution : Malaya, Thailand, Sumatra, Java. Burma (Lace 1922).

2. *Sessiles*

Bracts of the male spike cylindrical with margins not bent outwards, flowers included see also *G. contractum* Mfg and *G. latifolium* Bl. Species 11, varieties and forms 4; species only 2 within our area.

Key to the Species
(Male Plants) After Bharadwaja 1957.

1. Male spikes 5 × 0.7 cm; lamina coriaceous, brown in dried specimens, 18 × 8 cm .... 7. *G. macrostachyum*

1 . Male spike 1.5 × 0.35 cm; lamina fleshy, green, 13 × 5 cm .... 8. *G. microcarpum*

(Female Plants)

1. Female inflorescence 9 × 1 cm; fruiting axis up to 10 cm long; lamina coriaceous, 18 × 8 cm; elliptic .... 7. *G. macrostachyum*

1′. Female inflorescence 3 × 0.5 cm; fruiting axis up to 6 cm; lamina fleshy, 13 × 5 cm, oblong-elliptic or lanceolate .... 8. *G. microcarpum*

7. **Gnetum macrostachyum** Hook.f. FBI. 5: 642. 1890. (Pl. 37, Fig. 11)
Mfg. op. cit. 484; Bharadwaja op. cit. 419.

A woody liana. *Branches* lentcellose. *Leaves* leathery, lamina elliptic, acuminate up to 18 × 8 cm, base rounded; veins distinct below, secondary curved, 6-8 on either sides, distinctly joined, tertiaries reticulate, petiole 1 cm long. *Inflorescence* simple up to 5 cm × 7 mm; bracts cylindric, 3 mm high. Male *Flowers* club-shaped 1.5 mm long, intermixed with long dark hairs, sporophyll shortly exserted. Sporangia 2, ellipsoid; *Imperfect* ♀ *flowers* about 10, oblique, ovate, fleshy, slightly exserted. ♀ *Inflorescence* simple often cauliflorus up to 9 cm long. *Flowers* 8-10, globose, apiculate, surrounded by profuse hairs; perianth fleshy; outer integument chartaceous, smaller than perianth, inner integument prolonged into a tube, exserted

1.5 mm. 'Fruit' sessile, shining, ellipsoid, mucronate 2 × 1.2 cm, epicarp thin, fleshy, mesocarp coriaceous; endocarp shining.

Locality : BURMA, Tavoy, *Parker* (Herb DD).

Distribution : Malaya, Sumatra, Java and New Guinea.

8. **Gnetum microcarpum** Blume in Rumphia 4: 7, t. 175, Fig. 1, 1848; Mfg. op. cit. 485; Bharadwaja op. cit. 419; *G. neglectum* Kurz in Flora 55: 350. 1872; For. Fl. Brit. Burma 2: 946. 1877; Hook.f. FBI 5: 642. 1890; Bharadwaja op.cit. 420. (Pl. 38, Fig. 12)

A woody climber. *Branches* slender. *Leaves* coriaceous, fleshy, shining, turning black on drying, oblong-elliptic, or lanceolate acuminate, 13 × 5 cm, base rounded or cuneate, veins 7-9, curved towards the margin, petiole 6-8 mm. Male *Inflorescence* cauliflorus, peduncle up to 1.5 cm × 3.5 mm. Male *Flowers* 40-80, 1.5 × 5 mm; sporophyll 3.5 mm, sporangia 2, yellow. *Imperfect flowers* 20-30, ovate or fusiform 1.5-2 mm long. ♀ *inflorescence* cauliflorus simple, dense, spike 2.5 cm × 5 mm. ♀ *flowers* 8, 3 × 2 mm, ovate, acuminate, apex slightly curved, perianth fleshy, fibrous. *'Fruit'* red, smooth, ellipsoid, shining, acute up to 2 × 1.2 cm; *'epicarp'* thin, fleshy, sparingly fibrous; *'mesocarp',* hard coriaceous; *'endocarp'* papery or of tissue-like texture. *Seed* oblong up to 1 × 0.5 cm.

f. **silvestre** Ridley in *J. Roy. Asiat. Soc. Bengal* Calcutta 60: 62, 1911; Mfg. op. cit. 486.

Leaves : oblong, elliptic with a rounded base.

Locality : BURMA, Mergui *Griffith* 4972 (Herb. CAL).

Distribution : Malaya to Sumatra.

Reported from Malaya at c 2000 m on Mt Tahan.

Distribution of Gnetum in India and adjacent countries

Name	Number of species in our area.	Area of distribution and range of altitude.
Gnetum	8 species 4 varieties and two formas.	W. Ghats, Eastern coast and Ghats including Orissa, W. Bengal, Sikkim, Bangladesh, Arunachal Pradesh, Assam, Nagaland, Manipur, Mizoram, Andaman and Nicobar Islands and Burma. None in Sri Lanka.
G. gnemon		The tree form in our area is only found in the Gt. Nicobar Island where it was recently reported by Thothathri *et al* (1976). Elsewhere it is found in Indonesia which is adjacent to the Gt. Nicobar Island.
G. gnemon var. *brunonianum*		Assam (Golaghat), S. Mizoram Hills, Naga Hills, Burma (Kachin Hills, Mergui, Tavoy Dist., Tenasserim).
G. gnemon var. *griffithii*		Assam (Sibsagar), Naga Hills, Kungba 1500 m. Burma (Tavoy, Tenassrim River, S. Tenasserim, Nagawun, Chaung forest, Mergui) Ascends up to 1500 m in the rainforests in Naga Hills and Mizoram.
G. oblongum		Bangladesh (Chittagong, Demagiri), Burma (Tonkyeghat, Tavoy, Wazum Chaung, Mergui, Lemji).
G. ula		Maharashtra (Yellapur, Khandala Valley, Poona). Goa, Kanara, Karnataka (Coorg), Kerala (Nilambur, Quilon), Tamil Nadu (Nilgiris, Devamalai at 1300 m in Nilgiris, Pulni Hills, Chenath Nair Hills, Andhra Pradesh (Godavari Dist), Orissa (Ganjam Mahendragiri), S. Andamans.
G. montanum		Darjeeling, Sikkim at 1500 m, Bangladesh (Silhet) Assam, Orissa (Mayurbhanj), Burma (Katha Dist, Pegu, Tharrawaddy, Martaban. E. Nepal & C. Nepal.
G. montanum *f. megalocarpum*		Arunachal Pradesh (Dafla Hills).
G. contractum		Tamil Nadu (Conoor in Nilgiris), Kerala (Quilon) up to 5000 ft.

Name	Number of species in our area.	Area of distribution and range of altitude.
G. latifolium		
G. latifolium var. *macropodum*		Andamans (Port Blair, Mt. Harriat), S. Andamans. Nicobars (Kamorta).
G. latifolium var. *funiculare*		S. Andamans (Port Monat).
G. macrostachyum		Burma (Tavoy).
G. microcarpum f. *silvestre*		Burma (Mergui).

References

Arnold, C.A. 1948. Classification of the gymnosperms from the view point of palaeobotany. Bot. Gaz. 110: 2-12.

***Bertrand,** C.E. 1874. Anatomie comparee des tiges et des feuilles chez les Gnetaces et conifers *Ann. Sci. nat. (Bot)* 20: 5.

Bharadwaja, R.C. 1957. Genus *Gnetum* Linn in India, Pakistan and Burma. *J. Indian bot. Soc.* 36; 408-420.

Brandis, D. 1874. Forest Flora of N.W. and C. India.

Brandis, D. 1906. Indian Trees. London.

Cheng Ching-Yung 1978. *Ephedra, Gnetum* in Flora Reipublicae Popularis Sinicae, 7, Science Publications, Beijing, 544 pp.

Dutta, S.C. 1973. *An Introduction to Gymnosperms.* Asia Publishing House, Bombay.

Eames, A.J. 1952. Relationship of Ephedra. Phytomorph. 2: 79-100.

Fagerlind, F. 1947. Strobilus und Blute von *Gnetum* und die Moglichkeit, aus ihrer struktur den Blutenbau der Angiospermen zu deuten. Arkiv Bot. 33.A (8): 1-57.

Florin, R. 1934. Ubereinge neue oder wenig bekannte *Ephedra-* Arten der Sect. *Pseudo-Baccatae* Stapf. *K. Svensk. Vetensk-Akad.* Handl. 12: 1-44.

Foster A.S. & E.H. Gifford. 1959. *Comparative Morphology of Vascular Plants.* W.H. Freeman & Co. San Francisco pp. 320-442.

Gagnepain, F. 1946. **Bull. Soc. Bot. Fr. 93: 313-320.**

Hara, H. Stearn, W.T. & Williams, L.H.J. 1978. An Enumertion of the Flowering Plants of Nepal (Vol. I) British Museum London.

Hooker, J.D. 1890. Flora of British India Vol. 5.

Khoshoo, T.N. & M.R. Ahuja, 1963. *Welwitschia mirabilis. Chromosoma* 14: 522-533.

KURZ, S. 1877. Forest Flora of British Burma.

Lam, J. 1951. Dynamic Palaeontology. *3rd Congr. Strat. et Geol. Carb.* Heerlan : 385-394.

Madhulata, 1960. Morphology and embryology of *Gnetum gnemon* L. Part. I. Ph. D. thesis. Univ. Dehli.

Maheshwari, P. & V. Vasil (1974). Botanical Monograph No. 1. GNETUM. CSIR, NEW DELHI pp 1-142.

Markgraf, F. 1926. Gnetales. In Engler and Prantl, Die naturlichen Pflanzenfamilien. ed. 2, bd 13: 407-441.

Markgraf, F. 1930. Monographic der Gattung *Gnetum. Bull Jard. bot.* Buitenz. Ser. 3, 10: 407-511.

Markgraf, F. 1954. Gnetaceae. In *Flora Malesiana* 4: 336-347.

Matthew, K.M. & Rani, N. 1983. The Flora of Tamil Nadu Carnatic pt. 2 (Edit. K.M. Mathew). Tiruchirapathi.

Melchior, H. and E. Werdermann. 1954. A Engler's *Syllabus der Pflanzenfamilien,* Vol. I, *Bacterien-Gymnospermen,* Berlin.

Pant, D.D. 1957. The classification of Gymnospermous plants. *Palaeobot* 6: 65-70.

Pearson, H.H. W. 1929. Gnetales. Cambridge University Press London.

Rheede, H.V. 1688. *Hortus Malabaricus* 7.

Roxburgh, W. 1832. *Flora Indica,* Calcutta.

Sporne, K.R. 1965. *The Morphology of Gymnosperms.* Hutchinson & Co. (Publishers), London.

Thierot, J.W. 1980. In : Gymnosperms, The New Encyclopaedia Britannica 8: 523, 15th ed. Helen Hemingway Barton, London.

Thothathri, K., S.P. Bannerji, P.K. Hajra & C.D. Pal. 1973. Botanical Results of the Joint Scientific Expedition to the Great Nicobar Island Bull. Bot. Surv. India 15 (Nos. 3 & 4): 235-265.

Vasil, V. 1959. Morphology and embryology of *Gnetum ula* Phytomorphology 9: 167-214.

FAMILY 11. **EPHEDRACEAE** Dumortier *nom. cons.*

Dumortier, Anal. Fam. Pl. 2: 12. 1829. **EPHEDRA FAMILY**

Shrubs with one climber in our area. Much branched with opposite connate *leaves* reduced to scales, so that the stem performs the function of assimilation. *Flowers* unisexual with no trace of carpels in male or stamens in female; Male in spikes. Female in pairs or solitary, usually bracteate. The male has a perianth of two antero-posterior united leaves, beyond which the axis is prolonged and bears 2-8 sessile 2-locular anthers. The female has a tubular perianth and one erect orthotropus ovule with a long micropyle projecting at the top of the flower, the flower or flowers are enclosed by bracts which become red and fleshy after fertilization and enclose the *fruit.* The *seed* enclosed in the perianth, which becomes woody, and the fleshy bracts cover this again. Two cotyledons in the embryo; seed albuminous. The only genus is *Ephedra* distributed in warm temperate North and South America and Eurasia.

1. **Ephedra** (Teurn. ex L. gen. Pl. ed. 1 (1737) 312) L. Sp. Pl. ed. 1. (1753) 1040. FBI V. 640. 1890. Stapf. Monogr. Die Arlen der Gattung *Ephedra,* Wien, 1889. Brandis, Indian Trees 1906, 686. Markgraf. F. Gnetales. In Engler and Prantl, Die naturlichen Pflanzenfamilien ed. 2. 13: 407-441. 1926.

Rigid shrubs, erect, or decumbent, one species climbing, old stems often ribbed, branches green for several seasons, terete, striate, *leaves* mostly deciduous, opposite or whorled, mostly connate basally and usually reduced to membranous sheaths at the nodes, which often bear 2-4 narrow leafy blades, plants dioecious or rarely monoecious, *staminate flowers* in compact subglobose to oblong, compound, usually axillary strobili, the latter usually opposite or in whorls of 3 or 4 at twig nodes, each strobilus bearing 2-8 opposite pairs of green bracts along the primary axis, bearing solitary male flowers, each flower borne on a short secondary axis arising between each fertile pair of bracts, the flower (sporangiophore) composed of 2 thin, opposite scales (perianth of some authors) above which are 1-8 stames with sessile or filamented anthers, the anthers dehiscing by terminal pores. *Ovulate* flowers in an elongated acute strobilus, the strobili opposite or in whorls of 3-4 at branch nodes, each strobilus with usually many pairs of bracts, the lower pairs sterile, and bearing usually terminal naked ovule, the ovule enclosed by 2 integuments, the outer integument represented by 4 basally coherent bracts (sometimes referred as the perianth), the inner of 2 similar bracts that elongate considerably at time of pollination and style-like in appearance (tubullus of some authors). It has the function of a style. *Seed* with leathery integument, globose to cylindric, membranous or winged or forming a berry-like syncarp, often red in colour, cotyledons 2.

The genus is of wide sporadic distribution in the semi-arid tropics and subtropics of northern and southern hemispheres and represented by about 44 species. About 18 species occur in the Old World (France, Canary Islands, around the Mediterranean east to Iran, Pakistan, India, and China), none in Sri Lanka and about 24 in the New World with 13-16 in N America and 9-11 in S. America.

The plants are readily distinguished by their evergreen horsetail-like stems, usually open and straggling habit, and small deciduous cones borne in stem axils.

Eight taxa occur in our area. *E. foliata* Boiss. the only climber occurs in the Punjab, Sind and Rajasthan, *E. gerardiana* Wall. ex Stapf, in the semi-arid alpine and temperate Himalaya from Kashmir to Kumaon, *E. intermedia* Schr. ex Meyer in Kashmir, Himachal and Garhwal in Jaunsar, *E. nebrodensis* Tineo in Kashmir, *E. pachyclada* Boiss in Chitral, *E. saxatilis* (Stapf) Royle ex Florin in W. Himalaya and var. *sikkimensis* (Stapf) Florin in Sikkim, *E. regeliana* Florin is rare and is found in Ladakh. The most rare *E. przewalskii* Stapf of Central Asia occurs on Nanga Parbat in Kashmir reported by Carl Troll (1939) and in the bed of Shyok River in Ladakh.

Fossil types : Wilson (1969) found pollen grains of *Ephedripites* which is regarded to belong to the ancestral form of *Ephedra*. The pollen grains of *Ephedra* have been discovered from "early Tertiary horizons in Tasmania, Victoria and S. Australia, in the Eocene Green River Shales of Colorado and the Paleocene of Hanovar, Germany, and at Cretaceous horizons in Iraq, Venezuela and Long Island in New York (Andrews, 1961).

Cytology : By culturing the pollen grains, a study has been made of the number and morphology of chromosomes. Mehra (1950) recorded the haploid number of 7 for *E. foliata* and *E. gerardiana* and 14 for

E. intermedia and *E. saxatilis.* The two male nuclei are of the same size in *E. foliata* and may be equal or subequal in *E. intermedia* and *E. saxatilis.*

Habit : The plants consist in the main of switchy green shrubs with a lone climber, *E. foliata* Boiss., which is found in the semi arid parts of the Punjab, Sind and Rajasthan.

Leaf : The leaves are represented by dry scales, which are decussate in arrangement and fuse at the base to form a sheath. They have parallel venation.

Reproductive organs : The plants are dioecious, bisexual inflorescences have been occasionally reported. Pearson (1929) observed hermaphrodite flowers on the same plant. These are considered as monstrosities by Eames (1952).

Female strobilus : The fertile female shoot consists of a greatly reduced axis or sporophyll and usually a single ovule. The ovules are enclosed by envelopes in addition to the integument. The upper portion of the inner envelope is prolonged considerably to form the micropyle tube.

Seed : The seed is enclosed within a reddish fleshy scale (derived from the fusion of the subtending bracts of the female strobilus) and has no resting period. During germination, the cotyledons grow upward and perform photosynthesis even after the first scaly leaf is formed. There is a remarkable degree of uniformity in vegetative form, and in the structure of the flowers, and to a less extent in that of the inflorescences. This introduces a degree of difficulty in separating and grouping the species. The general habit is of little value as a systematic character, as in most species it is very much influenced by environment. The system in vogue relies for its main subdivision chiefly upon the characters of the bracts of the female inflorescence which are diagnostic. The genus is subdivided into 3 sections. Section II is not represented in our area.

Key to Sections/Tribes (Mostly after Pearson 1929)

Section I :

Alatae Stapf. Mature female spike dry; bracts more or less hardened along the midrib; otherwise membranous, produced laterally into wing-like expansions.

Tribe Tropidolepides Stapf. Mature bracts hardened along the midrib and at length rather thickened. *E. przewalskii* Stapf.

Section II :

Asarca Stapf. Mature female spike dry; bracts slightly hardened, scarcely membranous-winged.

(Taxa not in our area. N. America).

Section III :

Pseudobaccatae Stapf. Mature bracts of female spike often narrowly membranous-margined, at length becoming thick and fleshy.

Tribe Scandentes Stapf. Shrubs usually scandent or subscandent. Micropylar tube straight when young, usually also when mature. *E. foliata* Boiss.

Tribe Pachycladae Stapf. Shrubs usually less than one m. Branches very rigid, thick and erect. Micropylar tube contorted.

E. pachyclada Boiss., *E. intermedia* Schrenk et C.A. Mey var. *tibetica* Stapf.

Tribe Leptocladae Stapf. Low shrubs with slender rather rigid branches. Micropylar tube straight or contorted. *E. gerardiana* Wall. ex Stapf., *E. nebrodensis* Tin. var. *procera* Stapf., *E. saxatilis* (Stapf) Royle ex Florin, *E. saxatilis* (Stapf) Royle ex Florin var. *sikkimansis* (Stapf.) Florin, *E. regeliana* Florin.

The genus Ephedra consists mainly of swtichy green shrubs with one climber *E. foliata* which is found in Sind, Rajasthan and the Punjab. This species scrambles over other vegetation and the stems attain 8 cm in diameter. The stem is a ribbed phylloclade, which carries on the function of photosynthesis. The shoot apex shows a well defined tunica layar, which is considered advanced in the anatomical sense (Gifford 1943). Parker (1918) described the plant thus "it climbs over small trees and covers them with a dense mass of branchlets, similar to *Cuscuta reflexa* (Convolvulaceae) at a little distance but of a different colour. The plant is also not unlike *Calligonum polygonoides* (Polygonaceae) and *Periploca hydaspidis* (Asclepiadaceae). It is a climber that is unlikely to be mistaken for any other plant if the flowers are also examined." It is also separated from other Ephedras by its more slender branches.

Troll (1939) records a rare Ephedra *E. przewalskii* Stapf of Central Asia from Nanga Parbat and in the bed of Shyok River in Ladakh reported by Stewart (1972). This belongs to Section Alatae, Tribe Tropidolepides, diagnostic characters of which are given in the key above. This is the sole member of Alatae in our area, all the rest of the taxa belong to the Section Pseudobaccatae.

Key to Species

Climbers, branches slender, filiform .... 1. *E. foliata*

Not climbers, all shrubs. all the rest of the species.

Note : for determination of *E. przewalskii* turn to key to Sections/ tribes p. 136 and diagnosis/descriptions on p 141.

Tubillus twisted :

- Seeds 2-3 mm long; bracts of female inflorescence connate in the lower third .... 3. *E. pachyclada*
- Seeds 5-6 mm long; bracts of female inflorescence connate in the lower two-third .... 4. *E. intermedia*

Tubillus straight :

- Tubillus not more than 1 mm long:
  - Strobilus 10 mm long; seeds 6.5-7.5 mm long; tubillus 0.5 mm long .... 5. *E. saxatilis*
  - Strobilus 5-8 mm long; seeds 4-6 mm long, tubillus one mm long .... 6. *E. gerardiana*
- Tubillus longer, 1.5-3.5 mm long :
  - Branches very smooth; strobilus 5-7 mm long; seeds 4-7 mm long .... 7. *E. nebrodensis*
  - Branches smooth or scaberulous, strobilus 7-10 mm long; seeds 3.5-4.5 mm long .... 8. *E. regeliana*

Distribution of the various taxa

Tribe Tropidolepides

E. przewalskii Stapf : Nanga Parbat (Kashmir), along bed of Shyok-River in Ladakh.

Tribe Scandentes Stapf.

E. foliata Boiss : Baluchistan, Sind, Rajasthan, Punjab.

Tribe Pachycladae Stapf.

E. pachyclade Boiss : Chitral, Baluchistan, Kurram Valley (NWFP Pakistan).

E. intermedia Schrenk et C.A. Mey var *tibetica* Stapf Lahul, Spiti, Pangi, Rowli Chamba, Ladakh.

Tribe Leptocladae

E. regeliana Florin : Karokorum, Ladakh.

E. gerardiana Wall. ex Stapf Karakorum, Chitral, Gilgit, Baltistan, Hazara, Kahsmir, Ladakh, Zanskar, Kistwar, Chamba, Lahul, Garhwal, Kumaon, Badrinath.

E. saxatilis (Stapf) Royle ex Florin, Simla, Jaunsar (Garhwal) Badrinath, Kunawar, Kumaon, S. Tibet.

var. *sikkimensis* (Stapf) Florin : N. Sikkim (Chho Lhamo 17000′) S. Tibet.

E. nebrodensis Tin. var. *procera* (Fisch et Mey) Stapf, Baluchistan, Kashmir, Chamba, Lahul.

1. **Ephedra foliata** Boiss. et Kotschy Sched. Pers, Aust. 1845 et in Boiss Fl. Orient. 5: 716. 1885. J.D. Hooker, FBI 5: 863. 1890. Brandis Ind. Trees 686, 1906. Talbot 11 : 541, 1911. Parker 536, 1918. Bhandari Fl. Ind. Desert. 1978, p. 439. *E. peduncularis* Boiss Fl. Orient. 5: 717, 1885. *E. Alte* non C.A. Meyer Brandis For. Fl. 501, t. 69. 1874. (Pl. 39, Fig. 13)

Vern. *Kuchar* - Punjabi. *Lana, Suo-Phogara*-Rajasthan.

A tall much branched climbing shrub up to 8 cm in diameter. *Bark* brown, exfoliating in fibrous shreds; branches slender, drooping, usually whorled; branchlets smooth, striate, filiform, dark green or glaucescent, internodes 2.5-10 cm long.

Leaves : 2-3 at each node, shortly connate at the base, linear, setaceous, 2.5-3 cm long. *Flowers,* Male fls. in ovate, lateral or terminal often whorled and clustered spikelets, 1-3 together on 1-2 cm long pedicels. Flowers 6-24, bracts 1.5-2 mm apex ciliate. Staminal column exserted. Anthers 2-5.

Female fls. in 2-3 flowered cymes or fascicles pedicels 2-10 mm long, usually in pairs with 3 pairs of bracts, outer 1 × 1 mm long, margin scarious ciliate, the innermost 5 mm long, connate. Tubillus truncate or bilobed, straight, short, exserted from the obtusely trigonous integument. *Fruit* ovoid, 7-8 × 6.7 mm, succulent, red or milky-white when ripe, fleshy, semi-transparent. *Seeds* 2.67 × 2.25 mm planoconvex, truncate at apex, rounded at base, dorsally reeled; dark coloured.

Flowers : Feb. - March - April. Fruits ripen in May.

Distribution : Sind (Hills near Karachi - T. Cooke 1907), Baluchistan, Kurram Valley, N.W.F.P. Pakistan, Punjab Plains rising to 800 m in Salt Range, Rajasthan, Afghanistan, Arabia, Aden.

Specimens examined : PUNJAB, *N.C. Nair* 36544 (BSD), Faridkot, *N.C. Nair* 20095 (BSD), Laharu. Bhandari l.c. cites JODHPUR 1668, 1994; *Rolla* 67236 (BSI): BARMER 648, 649. JAISELMER 1080.

Uses : Stems and branches used for cleaning brass and metalic utensils. var *ciliata* (Mey) Stapf Monogr. Art. Gatt. Eph, Wien 1889.

Specimen examined : Changamanga Pakistan, without collector (DD). Det. R. Florin 1932. var. aitchisonii Stapf l.c. syn. E. ciliata Aitchison occurs in Kurram Valley. See Aitchison Fl. Kurram Valley, Afghanistan J. Linn. Soc. 18: 5-25-97, 1881.

2. **E. przewalskii** Stapf. in Danschr. Akad. Wiss. Wien (1889) 40. (Pl. 40, Fig. 14)

Diagnosis : Wings of bracts of mature female strobilus broad, laterally produced, 2-2.5 times broader than the hardened dorsal part, incurved at apex, narrowly oblong; seeds ovate, not attenuate in the neck, 4-5 mm long.

Description : Erect, suberect or ascending dioecious shrub, upto 1¼ m tall, branched from base, branches less divaricate. *Branches* rigid hard upto 2½ mm thick, terete, 20-25 cm long, internodes 3½ cm rarely 4 cm long, generally many (upto 12), arraanged in false whorls, rarely shortened, close set and sometimes circinate. *Bark* yellow green or turning brown, rough, thinly striate, pale yellow (reddish) brown, longitudinally or subreticulately cleft. Rhytidoma dark brown, full of cracks, not fibrous.

Leaves : in 3s or 2s, reduced at sheaths, 3 mm long, upto halfconnate, herbaceous green, membranous pale, sheath with teeth ovate or triangular, narrow, acute or becoming obtuse, often apiculate, narrowly membranous margined, on drying brownish, cleft.

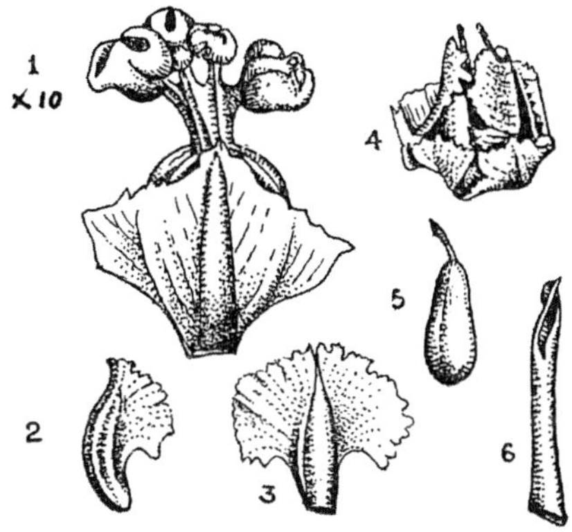

Ephedra przewalskii, 1 Male strobilus x 10 staminal column exerted. 5 seed oblong ovate. 6. Straight tubilus. From O. Stapf Die Arten den Gattang Ephedra Taf 1 (III).

Flowers : Male spike solitary, two or rarely three together at base of branches and in lower and intermediate nodes, sessile, spike ellipsoid or nearly spherical upto 4 mm long, whorls 3-flowered 4-6. Bracts yellow green rhomboid or rhomboid elliptic, 2 mm long and broad, base connate acute, margin broadly membranous hyaline, subentire. Perianth obovate cuneate, equalling or a little exceeding the bracts. Staminal column exserted, upto 3 mm long. Anther 5-8, shortly or very shortly stipitate.

Female strobilus solitary subsessile or rarely with upto 8 mm long elongate internode, ovate or subspherical, pairs of bracts or ternate whorls 5, laxly imbricate. Bracts free rounded or transversely broader, dorsal part narrowly oblong herbaceous, obtuse or emarginate, margins laterally produced in wing, broad membranous, hyaline, very minutely erose, below middle abruptly narrowed in a short broad claw.

Flowers two or three together. Outer integument ovate obtuse. *Tubillus* straight, 2½ mm long, exserted, equalling 1/3 of the ligulate limb.

Mature female strobilus 5-6 mm long and broad, bracts laxly imbricate, not divaricate, apex incurved, 5 mm long and broad. *Seeds* pale brown, oblong ovate, dorsally convex, facially plane, 3-4 mm long.

Distribution : Nanga Parbat in Kashmir (Carl Troll 1939), along bed of Shyok River in Ladakh and in Central Asia.

3. **E. pachyclada** Boiss. Fl. Orient. V, 713. 1885. Brandis Ind. Trees. 686, 1906. Hook. FBI V. 641. 1890. Bamber Pls. of the Punjab 645, 1916. Riedl in Fl. Iranica 4, 1963. Hara, Stearn, and Williams : Enum. Fl. Pls. Nepal 1: 23. 1978. (Pl. 41, Fig. 15)

A small gregarious shrub, usually less than one m high. *Branches* stout and rigid, erect, striate, scabrous. Internodes 2.5-6.5 cm long, 0.3-0.5 cm in diameter. *Flowers,* male spikes densely crowded, sessile. Anthers 6-8, sessile, crowded and large. Female flowers, bracts of female inflorescence connate in the lower third. Tubillus or micropylar tube curved. *Seeds* 2-3 mm long.

Flowers : May.

Distribution : Chitral, Baluchistan, Kakar Lora Valley 5500-6000′ Kurram Valley, Ladakh, Nepal, S. Iran, Afghanistan.

Specimens examined : Kurram Valley. *Harsukh* 14900 (DD). Det R. Florin 1932. LADAKH, Minu Pureng Valley, 3800 m, *U.C. Bhattacharyya* 41485 (BSD), rigid shrub on dry cliff with male cones.

Parker 1918 considered this as a synonym of *E. intermedia* Schr. et Mey. This is ruled out on the grounds that seeds of *E. pachyclada* are smaller, 2-3 mm long, in *E. intermedia* seeds are 5-6 mm long (See Key).

The plant is like *E. gerardiana* but is larger, more stout and rough. Used in tanning and the fruits are edible.

4. **E. intermedia** Schrenk et C.A. Meyer in Mem. Sci. Nat. Acad. Petersb. Ser. 6.5. 1845 et in Monog. Ephed. 1846; Hook.f. FBI: 5. 863. 1890. Brandis Ind. Trees 686, 1906; Parker 536, 1918; Bobrov in Fl. URSS 1: 198, 1934. Nair N.C. Fl. Bashah. Him. 5, 1977. Hara, Stearn, and William, Enum. Fl. Pls. Nepal 1: 23. 1978. Chital-Gilgit-*Sumani,* Kunawar-*Khanna.*

An erect 1.2 m tall, densely branched shrub, prostrate below with long lax branchlets. *Branchlets* glaucous; internodes usually 2.5-5 cm long, a few 1.2-1.7 cm or occasionally 6.5 cm long, 0.2-0.3 cm in diameter, striate, scabrid on the sides and rough to the touch when dry; leaf sheaths 3 mm long or less, 2-toothed. *Flowers* male spikes subglobose, numerous, in dense whorls at the nodes; flowers about 8, bracts broadly obovate, obtuse, connate, 2-3 mm long, staminal column shortly exserted, anthers 5-6, densely crowded. Female spikes, solitary, paired or whorled, 2-flowered; tubillus exserted, twisted, sharply bent, long. *Fruit* 7 mm long, ovoid, red.

Flowers : May - June.

Distribution : Kashmir, Himachal Pradesh, Tibet, Yarkand, *Henderson sn.*

Specimens examined : LADAKH, Dras Kargil Road, 3300 m, on rocky soil, fairly common, *Wadhwa* 58766 (BSD). HIMACHAL PRADESH, Khalsa, Fruits red. *Wadhwa* 58938 (BSD). Khorangi-Pangi, *Lace* 1012 (DD). Ralli (Bashahr), *Kartar Singh* 45 (DD). Pooh, Kinnaur Dist, Tubullus twisted, *Janardhanan* 47822 (BSD).

Note : Closely allied to *E. pachyclada.* Not synonymous with it as regarded by Parker (1918). See note at the end of E. pachyclada.

var. **tibetica** Stapf. Stapf, Monogr. Art. Gatt. Eph. Wien, 62, 1889. Hook.f. FBI 5: 863 (1890). Hara, Stern & Williams En. Fl. Pls. Nepal i: 23. 1978. (Pl. 42, fig. 16)

A low shrub, 1.15 m tall, erect or prostrate below with sometimes long lax branchlets, internodes elongate more slender rarely short thick, bark scabrous or scaberulous glaucous sometimes bluish.

Distribution : Afghanistan, Gilgit, Ladakh, Kishtwar, N. Tibet. Nepal (Stapf p. 63. cites Schlaginweit, Hook.f. et Thoms. Specimens with nos.).

Specimens examined : HIMACHAL PRADESH, Khorangi-Pangi 7000-9000′, *Lace* 1012 (DD). Det. R. Florin. Rowli 8300′. Chamba *Parker,* 21684 (DD). Det. Florin. NEPAL, *Shrestha* and *Bist* 2348, 2700-4000 m. Vide Hara p. 23 - specimen not seen by me, identity doubtful because of ? in Hara p. 23.

5. **E. saxatilis** (Stapf) Royle ex Florin in Svensk. Vet. Akad. Sci. Handl. Ser. III. xii No. 1, 21 (1933); Hara, Stearn, and Williams En. Fl. Pls. Nepal 1: 23, 1978; *E. gerardiana* (Wall.) ex Stapf. var. *saxatilis* Stapf. Arten der Gattung Ephedra. Wien. 1889, p. 76, vern, *Dewaria* - Garhwal. (Pl. 43, Fig. 17)

Shrubs usually up to 15-20 cm. *Branches* almost erect or rising in the shape of a curve. Branchlets smooth, rough in specimens from N.W. Himalaya. Internodes 1-4.5 cm, mostly 2-3 cm long, 1.5-2.5 mm in diameter, grooved in dry state. Lateral branches in the lower reaches of the plant, sometimes higher up, occur in bunches or seemingly verticillately. Higher up they are opposite or more frequently separately at the nodes. Basal internodes elongated and flat. *Leaves* 2-5 mm decussate, fused in pairs, forming sheaves up to 1/2 to 2/3 of their length, skin-like, turning brown and finally shrink. Open part of the leaves, triangular, truncated or pointed, awl-shaped, curved out-wardly. *Flowers :* The male strobili arise singly or in bundles of 2 or 3 in the axils of leaves. Ovoid, or spherical 4.0 mm long and contains 2-4 pairs of flowers. Bracts ovate, truncate or acuminate, 2-3 mm long, fused basally up to 1/4 of their length. The stalk of the microsporangia (anther case) is *c* 3 mm long projecting over the flower involucrum being a bunch of 6-8 stamens with sessile anthers. Female strobili ovoid with one or two seeds. Pedicels 6 mm long. Lower bracts ovate almost circular, connate up to 1/5 of their length. Inner bracts ovate, acuminate or truncate, connate up to 2/5 or 3/4 of their length. Micropylar tube straight 0.5 mm in length exserted. The ripe strobilus with pulpy bracts is *c* 8 mm long. *Seeds* one to two, 5-6.5 × 2.5-3.2 mm, ovoid, pointed, convex, distinctly rimmed, greyish-brown.

Flowering & Fruiting : Flowers May-June, *Fruits* June-July.

Ecology & Distribution : It prefers relatively more humid but rocky habitats. This is evidenced by its occurrence around Khadamba Peak,

Mundali Road (Chakrata Division) on hill sides. The centre of its area lies in Jaunsar (Garhwal) and Kumaon. Most of the specimens in DD identified as *E. gerardiana* by early botanists turned out to be *E. saxatilis* after critical examination of DD sheets by Florin in 1932.

Distribution : N.W. Himalaya, Simla (Collet), Jaunsar, Mandali, Deoban, Almora Distt.

Specimens examined : GARHWAL, Chakrata-Mundali Road 8800′, *K.C. Sahni* 24178 (DD). Deoban-Mandali 9000′. *C.E. Parkinson* 7077 (DD), Mandali-Deoban 9000′ *J.F. Duthie* 2191 (DD) Karamba *J.S. Gamble* 22818 (DD).

var. **sikkimensis** (Stapf) Florin in Svensk. Vet-Akad. Sci. Handl. Ser. III. XII. No. 1 28 (1933). Hara, Stearn & Williams En. Fl. Pls. Nepal 1: 23. 1978. (Pl. 44, Fig. 18)

E. gerardiana var. *wallichii* Stapf in Denkschr. Math.)- nat. Cl. K. Akad. Wiss. Wien. LVI, 1889, p. 75 (partim). Syn. *Ephedra gerardiana* Wall. var. *sikkimensis* Stapf in Denkschr. Math. nat. Cl. K. Akad. Wiss. Wien, CVI, 1889, p. 76. *Ephedra vulgaris* auct non Rich. Hk.f. FBI 5: 640. (1890). This differs from the type *E. saxatilis* in its habit, it has a rounded crown (see Plate), smaller number of flowers in the male cone. (Pl. 44, Fig. 18).

Distribution : Semi-arid N Sikkim up to 17000′, Lachen (upto 3950 m).. S. Tibet, Chumbi valley, Nepal (Stainton 1670. Stainton For. Nepal 1972), Bhutan.

Specimens examined : N. Sikkim, Chho Lhamo 17000′ *K.C. Sahni* 83(DD): Llonakh, N. Sikkim, 14,500′ *K.C. Sahni* 62 (DD). *Relon & Romoo* 5403 (DD) without locality. *T. Anderson* 1522 (DD) without locality. The semi-arid Chho Lhamo plateau with the Chho Lhamo Lake is the source of the Tista River.

6. **Ephedra gerardiana** Wall. *ex* Stapf. (Numer List : 207 n. 6048 (1832), nom. nud.) *ex* Stapf in Akad. Wiss. Wien. Math. Naturwise. Kl. Denkschr. 56 (2): 75, t. 3, f. 18 (1889) Brandis Ind. Trees 686, 1906. Parker For. Fl. Punj. 535, 1918. Osmaston For. Fl. Kum. 559, 1927. Florin in Kungl. Svenska

GERAD'S EPHEDRA (Pl. 45, Fig. 19)

Vetenskdpasakad. Handl., Ser. 3, 12(1): 21, t. 3 (1933). Nair Fl. Bash. Him. 4. 1977. Hara, Stearn & Williams Enum. Fl. Pls. Nepal 1: 23, 1978. *E. gerardiana* var. *wallichii* Stapf in Denkschr. Math.-nat Cl.K. Akad. Wiss. Wien, LVI, 1889. *E. distachaya* Linn. Boiss. Fl. Orient. V. 713. 1885. Vern. *Tutgantha.* Jaunsar.

A rigid dense tufted shrub 15-120 cm high. *Setm* woody gnarled. *Branches* and branchlets dark green, erect or arcuately ascending. Internodes 1.3-3.8 cm long, 1.8-2.5 mm in diameter, striate, smooth or slightly scabrid on the ridges; leaf-sheaths 2.5 mm long. 2-toothed. *Male spikes* ovate, 1-3 together at the nodes. *Flowers* 4-8, bracts, obtuse, connate, 1.9-2.5 mm long; staminal column exserted, anthers 5-8. *Female spikes* usually solitary 1-2 flowered; tubillus exserted, straight. *Fruits* 8-10 mm long, clustered, ovoid, red. *Seeds* 2, black 4-6 mm long.

Flowers : May-July. *Fruits :* July-Oct.

Distribution : Occurs along the inner Himalayan ranges at 2000-4500 m Very common in inner dry ranges boardering Tibet on exposed shingly slopes. Karakorum C.B. Cl. 1876, N.W. Himalaya, Chitral, Baltistan, Dras, Sonamarg, Kishanganga, Chamba, Lahul, Kumaon, Milam (Florin 1933, p. 3).

Specimens examined : HIMACHAL PRADESH. U.C. *Bhattacharyya* 49300 (BSD) Spiti 3800 m, open stony slopes with female spikes. *N.C. Nair* 16628 (BSD) Lahul 3400 m. *K.P. Janardhanan* 46701 (BSD). U.P. Badrinath. *B.D. Naithani* 37396 (BSD) Gaumukh. *M.A. Rau* 8684 (BSD). Kedarnath.

Uses : Browsed by goats. Fruits very sweet edible. According to Florin (1933) *E. gerardiana,* which is adapted to relatively dry habitats, has its western limit in N.W. Afghanistan and its northern limit in the border region between the N.W. Himalaya and the Great Pamirs. Its vertical distribution is considerable 1450 m in Jhelum valley Kashmir to 5450 m in Karakorum.

var. **sikkimensis** Stapf in Akad. Wiss. Wien Math. Naturwiss Kl. Denkschr 56(2): 76 (1889). Banerji in J. Bombay Nat. Hist. Soc. 53: 153 (1955). Hara Fl. E. Him. 2: 13 (1971). Lancaster Pl. Hunt. Nepal 69, 119 (1981).

According to Grienson & Long 1983 p. 56 Fl. of Bhutan Vol. I, pt. The E. Himalaya plants are *E. gerardian* Stapf var. *sikkimensis* Stapf.

Erect, robust but soft, occurs in Sikkim and Nepal.

Uses : The rhizomes have large knots of the size of a football used as fuel. Suitable for the rock garden, with its ornamental red-current like berries.

Note : According to Florin (1933) this variety is the same as *E. saxatilis* var. *sikkimensis* (Stapf) Florin. I am in agreement with him. The only difference that is apparent is that this variety is described as robust but soft. *E. saxatilis* var. *sikkimensis* is robust but not soft. This is due to habit factor.

7. **Ephedra nebrodensis** Tineo in Guss Fl. Sic. Syn ii 638, 1832: Hook.f. FBI, V: 864, 1890: Stapf Arten der Gatt. Eph. Wiss. Wien 1889. 77. t. 3, xx f. 1-7. Brandis Ind. Trees, 686, 1906; Bor. Man. Ind. For. Bot. 25, 1953 Vern *Omah, 'Umah* - Baluchistan. (Pl. 46, Fig. 20)

An upright, densely branched shrub, upto 1.8 m high. *Branches* very smooth slender, rigid. Internodes smooth, slender 1.25-2.5 cm long, c. 2 mm in diameter. *Male spikelets* few flowered, solitary or 2-3 together, sessile crowded. *Female* cones 1-flowered, short pedunculate, innermost bracts connate for one-third of their length. Tubillus straight, 1.5-3.5 mm long, strobilus 5-7 mm long. *Fruit* globose 8 mm in diameter. *Seeds* 4-7 mm long.

Distribution : Baluchistan - Juniper tracts 7-10,000′. Baltistan. Lahul. (Brandis p. 687).

var. **procera** (Fisch et Mey) Stapf.

Branchlets perfectly smooth, semi-mature female cones narrower longer, seeds elongate-ovate.

Distribution : Lahul, Kyeling (Florin p. 81), Western Tibet, Baluchistan, Afghanistan and eastwards to Greece.

Specimens examined : Elliot 8504 (DD), Baluchistan (*E. nebrodensis* Tineo var. *procera* Fisch et Mey, Det R. Florin 1932).

8. **E. regaliana** Florin in Svensk Vet. Akad. Handb. ser. 3: 12(1): 17. 1933. Stockholm. syn. *Ephedra monosperma* Gmel. *E. disperma* Rgl. in Acta Hort. Petrop., vi: 2, 1882 p. 479, *E. monosperma* Stapf in Denkschr. Math. nat. Cl. d. K. Akat. Wiss. Wien, LVI, 1889, p. 73 (partim) (non Gmel apud C.A. Mey); *E. gerardiana* Stapf in Denkschr. Math-nat. Cl. d.k. Adak. Wiss. Wien, LVI, 1889, p. 75 (partim) (non Wall.) (Pl. 47, Fig. 21)

Etymology : The specific epithet is after Eduard von Regel of Leningrad Herbarium.

A very dwarf dioecious shrub, rarely exceeding 10 cm, much branched from the base. *Branchlets* erect, archedly ascending, slender, 0.7-1.5 mm thick, internodes short, 3.5-25 mm; usually 10-15 mm long, densely fascicled. *Bark* greyish green, scabrous to scaberulous, very thinly striate. *Leaves* opposite, herbaceous, membranous margined, soon turning pale for the major part, whitish membranous, brownish at base, 1/2-2/3rd part connate, 1-2.5 mm long, sheath with triangular teeth, short or more or less elongated, obtuse or acute. *Male strobili* single or 2-3 together, sessile or pedunculate, peduncle rarely up to 1.5 cm long. Strobilus globose or ovoid, up to 5 mm long. *Flowers* in 4-7 pairs. Bracts suborbicular-ovate, obtuse or subacute, membranous margined, connate up to 1/3, 2 mm long. Perianth obovate or rounded-obovate, equalling or somewhat exceeding the bract. Synangiophore entire or rarely a little divided exserted, *c* 3 mm long. Synangia 6-8 sessile or rarely short stipitate, glomerate. *Female* strobili 2 flowered (or 1-flowered), or axillary ones sessile or pedunculate (peduncle up to 2 cm long), usually solitary, or terminal, pairs of bracts 3 or rarely 2. Lower bracts up to 1/3 connate, broadly ovate, membranous margined, acute or subacute, intermediate ones 1/2-2/3 connate, broadly ovate, or suborbicular ovate, obtuse or subacute, equalling 1/2-3/4 length of upper ones, upper bracts up to 5/6 or a little beyond, connate lobes membranous margined. Flowers 2, opposite, shorter than bracts. Tubillus straight or more or less contorted, exserted, 1.5-2.5 mm long, limb ligulate margin erose. Mature strobilus globose, 7-10 mm long, bracts red, fleshy. *Seeds* included, 3.5-4.5 mm × 1.7-2 mm, dark brown, shining, ovate or oblong, obtuse, dorsally convex and faintly keeled, face subconcave.

Habitat in Central Asia.

Distribution : Its southern limit is in Ladakh. Karakorum 1892 *W.M. Conway.* 1876, C.B. Cl. N.W. Himalaya. Baltistan, Iskardo 1847. *J.E. Winterbottom.* Leh (Ladakh) 1857- *Schlaginweit* Pamirs, Tian Schen, Ala-tau, its northern boundary is in Tabagatai (Florin 1933) in Central Asia.

Ecology : On gravel, rocks, between stones in river valleys, at lakes and in dry hilly areas. The limit of its altitude is, c 4000 m (Florin 1933).

Specimens examined ; Dr. *Giles* s.n. (DD) Gilgit Expedition. Florin (1933) considers *E. gerardiana* (Wall.) Stapf var. *wallichii* to belong to *E. regaliana* Florin which has its southern limit in Ladakh.

This taxon is extremly rare in Indian herbaria. There is only one specimen *Giles* s.n. DD identified by Florin in 1932, none in BSD and CAL.

Ephedrine, a drug known to the Chinese for the past 5000 years, is the principal alkaloidal constituent of the plant *Ephedra sinica* Stapf, and is of great importance in the treatment of asthma and hay fever. Associated with it in the plant is its isomere pseudo-ephedrine. The bulk of the world's supply of this drug formerly came from China, but the Japanese war and internal unrest interfered with the production of the drug, so that it became imperative to look for alternative sources of supply. The chemists of the Forest Research Institute turned their attention to the members of the genus common in India, with interesting results. It now appears that *E. gerardiana* and *E. nebrodensis* contain ephedrine in quantities equal to or even superior to those found in *E. sinica.* Another species, *E. intermedia* also contains ephedrine but the main alkaloidal content is pseudoephedrine, which, however, is also valued as a drug. (Ghose Gupta and Krishna 1929). According to Singh (1950) the extraction of *Ephedra* from Lahul could be a commercial success provided the concentrates could be prepared at Kyeling itself and then sent to Manali in order to reduce the cost of transport.

The distribution list shows plainly that *Ephedra* is a plant of inhospitable arid regions. The plant appears to be able to stand long periods of drought and extremes of heat and cold. There seems to be some correlation between the annual rainfall and the amount of alkaloids present. A high ephedrine content is associated with a low rainfall and vice versa. The content of the drug is also said to be highest in the autumn. It has been suggested that probably *E. intermedia* Schrenk et C.A. Mey is the famous 'soma' plant of the vedic sacrificial rites, although the true 'soma' is still a matter of conjecture cf. *Sacrocostemma acidium* (Roxb.) Voigt.

In certain areas in the Himalayas *Ephedra* spp. constitute the only vegetation over large areas and the plants must be of considerable value in stabilising the soil on steep hill slopes. The practice, therefore, of pulling up this plant entire when harvesting, instead of leaving the root in the soil, is very much to be deprecated. Such practices only lead to erosion and eventually to the deterioration of the soil to an extent that it will not support any vegetation.

Discussion

Florin (1933) considers *E. gerardiana* (Wall.) Stapf var. *wallichii* to belong to *E. regeliana* Florin which has its southern limit in Ladakh. Areas of occurrence of the two are also similar viz. mostly Karakorum. Hara, Stearn & Williams (1978) treat *E. gerardiana* var. *wallichii* as under.

E. gerardiana Wall. (nom. nud.) ex Stapf in Akad Wiss Wien. Math. Naturwiss Kl. Denkschr 56(2): 75 t.3.f. 18 (1889). Hara, Stearn & Williams En. Fl. Pls. Nepal. 1: 23. 1978. *E. gerardiana* var. *wallichii* in Densksch Math. Nat. Cl. K.A. Akad. Wiss. Wien, LVI, 1889 p. 75. In other words Hara *et al.* merge it with *E. gerardiana* and do not recognize var. *wallichii.*

I agree with Florin (l.c.) where he observes.

"Certain Ephedrae from the N.W. Himalayas which are included in the species of *E. gerardiana* (var. *wallichii*) belong to *E. regaliana* newly described by me in this treatise which has its southern limit in Ladakh. Moreover, I conceived var. *saxatilis* introduced by Stapf, as a species of its own and added var. *sikkimensis* to it rather than to *E. gerardiana.* I believe to have reached a more natural classification than that proposed by Stapf in 1889".

Florin 1933 regards *E. gerardiana* Wall. ex Stapf var. *sikkimensis* Stapf a synonym of *E. saxatilis* (Stapf) Royle ex Florin I agree with Florin and not with Hara et al (1978) and Lancaster (1981) who call it *E. gerardiana* var. *sikkimensis* Stapf.

Stapf (1889, p. 77) was of the opinion that *E. saxatilis* shows certain relations with *E. gerardiana*. The form occurring in Sikkim he regarded as more independent and called it *E. gerardiana* var. *sikkimensis*. Florin (1933) is of the opinion that it would be best to consider *E. saxatilis* in contrast to *E. gerardiana* as a separate species. His experience is based on copious material. After intensive field work in N. Sikkim in the Chho Lhamo area at 17,000′ and in Llonakh at 14,500′, I came across abundant plants of *E. saxatilis* var. *sikkimensis* with upright stems and distinct rounded crowns (see sketch based on *Sahni* 62 (DD from Chho Lhama) and am in agreement with him and not with Hara, Stearn & Williams (1978) and Lancaster (1981) who consider this Sikkim plant a variety of *E. gerardiana*.

Florin l.c. has rightly concluded "Supposing *E. saxatilis* and *E. gerardiana* belong to one and the same species than one has to include in *E. gerardiana* not only the Sikkim plant, but also the Yunnan form as well as *E. minuta* from Szechuan, and one has to distinguish as many varieties within that multiforms species. But to me it looks such a solution is not satisfactory. Since the most western and the most north-eastern forms are relatively independent compared with the three others. I prefer to distinguish 3 species which can be considered closely related particularly because of the characteristics of their seeds and which form a natural group of species together with *E. likiangensis* Florin".

E. saxatilis is now clearly distinguished from *E. gerardiana* by the following characteristics: appearance and size of the plants, stronger branches, longer internodes, the deviating nature of knoblet formation at the surface of the branches, longer male strobilis, the larger number of flower pairs and the bigger bracts in those strobili, larger female strobili a shorter micropyle tube and longer seeds.

Another pair of confused taxa are *E. intermedia* and *E. pachyaclada*. Parker (1918) considered *E. pachyaclada* Boiss., a synonym of *E. intermedia* Schr et Mey. This is no longer correct because critical examination of the two taxa shows that seeds of *E. pachyaclada* are smaller 2-3 mm long, in *E. intermedia* they are 5-6 mm long. Bracts of the female inflorescence are connate in the lower third in the former, while in the latter they are connate in the lower two thirds.

Habitat : *E. regeliana* grows on gravel, rocks and between stones in river valleys at lakes and in dry hills. The limit of its altitude is around 4000 m. Its northern boundary is in Tabagatui. Otherwise it occurs in Ala-tau, Tian-Schan the Pamirs, Karakorum up to the N.W. Himalaya.

E. gerardiana is adapted to relatively dry habitats, its western limit is in N.W. Afghanistan and its northern limit is in the border region between the N.W. Himalaya and the Great Pamirs. Its eastern limit is Almora district. It prefers rocky habitats. Its vertical occurrence is considerable from the Jhelum valley in Kashmir at 1450 m to 5450 m in the Karakorums. *E. saxatilis* prefers relatively humid and rocky habitat. The centre of its area lies in Jaunsar in Garhwal and in Kumaon. It is interesting it has also been found much further east on Khemba La near Lhasa. Its range of altitude is 1970-4850 m. *E. saxatilis* var. *sikkimensis* grows at altitude of 3650-5000 m. in N. Sikkim and in the adjacent rainy parts of southern Tibet.

Bibliography (Ephedra).

Andrews, H.N. 1961. *Studies in Palaeobotany.* John Wiley & Sons, New York.

Boissier, E. 1885. *Flora Orientalis* V.

Brandis, D. 1874. Forest Flora of North Western and Central India.

Brandis, D. 1906. *Indian Trees.* London.

Cheng Ching-Yung 1978. *Ephedra-Gnetum* in Fl. Reipublicae Popularis Sinicae, 7, Science Publication Beijing, 544 pp.

Cooke, T. 1906. *The Flora of the Presidency of Bombay.*

Eames, A.J. 1952. Relationship of the Ephedrales. *Phytomorpholgy* 2: 79-100.

Florin, R. 1933. Uber einige neue oder wenig bekannte asiatische *Ephedra* Arten der sect *pseudo-baccatae* Stapf Kungl Svenska Vetenskapsakd Ser 3 12(1): 1-44, t. 1-5. Stockholm.

Ghose, T, Gupta, B.L. and Krishna, S. 1929. *Indian For.* LV: 215.

Gifford, E.M. 1943. The structure and development of shoot apex of *Ephedra altissima* Desf. Bull. *Torrey bot. Cl.* 70: 15-25.

Hara, H, Stearn, W.T. & Williams, L.H.J. 1978. *An Enumeration of the Flowering Plants of Nepal* Vol. I. Brit. Museum, London.

Hooker, J.D. 1880. *Flora of British-India* Vol. 5. 640-641, 863-864.

Lancaster, Roy 1981. *Plant Hunting in Nepal.* Vikas Publishing House, New Delhi.

Markgraf, 1926. Gnetales In Engler and Prantl, Die naturlichen Pflanzenfamilien ed. 2.

Mehra, P.N. 1950. A study of the Karyotypes and the occurrence of diploid gametophytes in some species of the genus *Ephedra. Proc. nat. Acad. Sci. India* 16; 259-286.

Meyer, C.A. 1846. Versuch einer Monographie der Gattung Ephedra.

Nair, N.C. 1977. *Flora of Bashahr Himalaya.* International Bioscience Publishers, Hissar.

Osmaston, A.E. 1927. *Forest Flora for Kumaon.*

Parker, R.N. 1918. A *Forest Flora for the Punjab with Hazara and Delhi.*

Pearson, H.H.W. 1929. *Gnetales.* Cambridge University Press, London.

Riedl, H. 1963. In *Flora Iranica.*

Royle, J.F. 1839. Illustrations of the Botany and other branches of the Natural History of the Himalayan Mountains and of the Flora of Cashmere. London.

Singh, D.P. 1950. Indian Ephedras and their supply. *Indian For.* 76: 288-289.

Stainton, J.D.A. 1972. *Forests of Nepal.* Hafner Publishing Company, New York.

Stapf, O. 1889. Monogr. Arten Der Gattung Ephedra, Wien.

Stewart, R.R. 1972. An Annotated Catalogue of the Vascular Plants of West Pakistan and Kashmir. Fakhri Printing Press, Karachi.

Strachey, R. 1906. A catalogue of Kumaon Plants.

Talbot, W.A. 1911. *Forest Flora of the Bombay Presidency and Sind.* Vol. 2.

Troll, Carl 1939. Das Pflanzenkleid des Nanga Parbat Begleitworte Zur Vegetationskarte der Nanga Parbat Gruppe Sonder-druck aus Wissenchaftliche Veroffentlichungen des Deutschen Kunde Zu Leipzig, N.F. 7

Wilson, L.R. 1959. Geological History of the Gnetales *Okla. Geol. Notes* 19: 35-40.

Bibliography (Gymnospermae)

The following reference sources are acknowledged.

Alvin, K.A. 1981. *The Oxford Encyclopedia of Trees of the World* (Ed. Bayard Hora) Oxford University Press, Oxford.

Andrews, H.N. 1961. *Studies in Palaeobotany.* John Wiley & Sons, New York.

Arnold, C.A. 1948. Classification of the gymnosperms from the view point of palaeobotany. *Bot. Gaz.* 110: 2612.

Banerji, N.L. 1954. Observation on the distribution of Gymnosperms in Eastern Nepal. *J. Bombay. nat. Hist. Soc.;* 51(1); 156, 59.

Becking, J.H. 1950. Der Anbau von *Pinus merkusii* in den Tropen. Schweizerischne Zeitschrift fur Forstwesen Journ Forestier Suisse, 101 Jahrgang, Mai.

Beisnner, L. 1930. *Handbuch der Nadebholzkunde* (Revised Fitschen, J.) Berlin.

Bentham, G. and Hooker, J.D. 1862-83. *Genera Plantarum.* Kew.

Bertrand, C.E. 1874. Anatomie comparce des tiges et des feuiles chezles Gnetaceae et conifers *Ann. Sci. nat. (Bot)* 20: 5.

Bharadwaja, R.C. 1957. Genus *Gnetum* Linn in India, Pakistan and Burma. *J. Indian bot. Soc.* 36: 408-420.

Biswas, K. 1933. The distribution of Wild Gymnosperms in the Indian Empire. *J. Indian bot. Soc.,* 12(1): 24 47.

Boissier, E. 1885. *Flora Orientalis* V.

Bor, N.L. 1935. Conifers of the Balipara Frontier Tract. *Indian For.* 61. (5): 313-19.

______, 1953. *A Manual of Indian Forest Botany.* Oxford Univ. Press

Brandis, Sir Dietrich 1906. *Indian Trees.* Archibald Constable & Co. Ltd., London.

Burkill, I.H. 1933, 1935. *A. Dictionary of the Economic Products of the Malay Peninsula.* **1, 2.** London.

Carriere, E.A. 1867. Traite genera des Coniferes, ed. 2.

Chamberlain, C.J. 1935. *Gymnosperms Structure and Evolution.* Chicago.

Collier, J.V. 1924. The eastern limit of the natural distribution of deodar. *Indian For.* L: 108-9.

Cowan, A.M. & Cowan, J.M. 1929. *The Trees of Northern Bengal,* Calcutta.

Critchfield, W.B. & Little, E.L. 1966. Geographic distribution of the pines of the World. USDA Misc. Publ. No. 991. Washington, D.C

Dallimore, W & Jackson, A.B. (4th edn. revised Harrison, R.G.) 1966. *A Handbook of Coniferae & Ginkgoaceae.* Edward Arnold, London.

Darlington, C.D. & Wylie A.P. 1955. *Chromosome Atlas of Flowering Plants* Aberdeen.

Dogra, P.D. 1964. Gymnosperms of India II. Chilgoza Pine (*Pinus gerardiana* Wall) *Bull. Nat. Bot. Gdn.* Lucknow.

______, 1985. Conifers of India and their wild gene resources in relation to Tree Breeding. *Indian For* Nov. 85, special issue pp. 935-55.

Dutta, S.C. 1973. *An Introduction to Gymnosperms.* Asia Publishing House, Bombay.

Eichler, W.A. 1883. *Syllabus der vorlesungen,* ed 3.

Engler, A. 1897. In Engler & Prantl *Die Naturlichen Pflanzenfamilien* Nachtrag Zu Teilin II-IV.

______ & Prantl. K. 1926. *Die naturlichen Pflanzenfam.* **13** edn 2. Gymnospermae 6, Pilger, R. Leipzig.

______ & Melchior *et al* 12th ed. 1954, 1964. Syllabus der Pflan. Bros Borntraeger, Berlin.

Ferguson, D.K. 1985. A new species of *Amentotaxus* (Taxaceae) from northeastern India. *Kew Bull.* 40(1): 115-119.

Fletcher, H.R. 1975. *A Quest of Flowers : The Plant Explorations of Frank Ludlow and George Sheriff.* Edinburgh University Press.

Florin, R. 1930. Die Koniferengattung *Libocedrus* Endl. In *Ostasein Svensk* bot. Tidskr. Bd. 24, H, 1.

———, 1948. On the Morphology and relationship of the *Taxaceae Bot. Gaz.,* 110(1): 31-39.

———, 1951. Evolution in *Cordaites* and *Conifers. Acta Horti Bergiani* Band 15, No. 11, Uppsala.

———, A Century of Progress in the Natural Sciences 1853-1953. (1955) 323-403, California Academy & Sciences.

———, 1956. Nomenclatural notes on genera of living gymnosperms. Taxon 5(8): 188-92.

———, 1958. *Acta Horti Bergiani* 17 (No II), 403-11.

———, 1963. The Distribution of Conifer and Taxad Genera in Time and Space. *Acta Horti Bergiani* Band 20(4): 121-311. Uppsala.

Fosberg, F.R. 1951. The Maldive Islands, Indian Ocean. (Atoll Res. Bull. 58) Washington.

Franco, J. do A 1969. On Himalayan - Chinese Cypresses Port. *Acta* Biol. Ser 139: 183-95.

Gagnepain, F. 1946. Bull. Soc. Bot. Fr. 93: 313-320.

Gamble, J.S. 1922. *A Manual of Indian Timbers.* London.

Gelderen. D.M. Van & Hoey Smith, J.R.P. van 1986. *Conifers,* B.T. Batsford Ltd. London.

Grierson, A.J.C. & Long, D.G. 1983. *Flora of Bhutan* Vol. I Part I Royal Botanic Garden, Edinburgh.

——— & Page, C.N. 1980. Notes relating to the Flora of Bhutan : III Pinus bhutanica : a new 5-needle pine from Bhutan and India. *Notes Roy. Bot. Gard. Edinburgh.* 38: 297-310.

Hara, H. 1966. *The Flora of Eastern Himalaya.* University of Tokyo Press, Japan. 2, 3 by Hara & Ohasi, 1971, 1975.

Hara, H, Stearn W.T. & Williams, L.H.J. 1978. *An Enumeration of the Flowering Plants of Nepal.* Vol. I, British Museum London.

Haridasan, K. 1988. Amentotaxus (Taxaceae): A rare Gymnosperm from Arunachal Pradesh. *Indian For.* 114(12): 868-870.

Hirase, S. 1895-1896. Etudes sur la fecondation et l' embryogenic du *Ginkgo biloba. J. Coll. Sci. Imp. Univ. Tokyo.* 8: 310-22. 1895. (Announcing Spermatozoid of *Ginkgo*). *Bot. Mag. Tokyo,* 10: 171-72, 1896.

Hirase, S. 1898. Etudes sur la fecondation et l' embryogenic du *Ginkgo biloba,,* II, *J. Coll. Sci. Imp. Univ., Tokyo* 12: 103-149.

Hofmeister, W. 1851. *Vergleichende Huntersuchungen der hoherer Crptogamen und der Samenbildung der* Coniferen, Leipzig (Engl. Translation by Currey, F. 1862.

Hooker, J.D. 1883. Flora of British India Vol. V (Secretary of State for India, London.)

———, 1906. A Sketch of the Flora of British India. pp. 5-60. Clarendon Press Oxford.

Hora, Bayard 1981. *The Oxford Encyclopedia of Trees of the World.* Oxford University Press, Oxford.

Hu, H.H. & Cheng W.C. 1948. On the new family *Metasequoiaceae* and on *Metasequoia glyptostroboides,* a living species of the genus *Metasequoia* found in Szechuan and Hupeh. *Bull. Fan. Mem. Inst. Biol. Peking, N.S.,* 1(2): 153-61.

Hu-Lin, Li 1953. A reclassification of Libocedrus and Cupressaceae. *J. Arnold Arbor.,* **34,** 17-36.

Hui-Yen, Chow 1974 June. International Wood Collecters Society 27(6).

Hunt, D.R. 1968. *Amentotaxus* from Assam. *Kew Bull.* 22(3): 402.

Jingwei, Z. 1982. *The Alpine Plants of China.* Science Press, Beijing, China. Gardon and Breach, Science Publishers Inc. New York.

Kanjilal, U.N., *et al* 1940. *Flora of Assam* 4: 332-347. Shillong.

Kermode, C.W.D. 1945. Mine eyes unto the hills. *Indian For.* 71: 1-13.

Kingdon-Ward, F. 1916. *Trans. bot. Soc. Edin.* XXVII. 49.

———, 1952. Plant Hunting in Assam. *J. Roy. Hort. Soc.* LXXVII (pt 6) 205-214.

Komarov, V.L. 1924. Not Syst Herb. Hort. Petrop V: 26.

Kurz, S. 1877. *Forest Flora of British Burma.* Supdt. Govt. Printing, Calcutta.

Lace, J.H. 1922. List of Trees, shrubs and Principal Climbers recorded from Burma. ed 2. rev. A. Rodger, Rangoon.

Lam, J. 190. Stachyospory and phyllospory as factors in the natural system in the Cormophyta. *Svensk. Bot Tidskr.,* 44: 517-34.

Lam, J. 1951. Dynamic Palaeontology 3rd Cong. Strat. et Geol. Heerlan 385-394.

Lambert, W.J. 1924. The eastern limit of the distribution of deodar. *Indian For.* L. : 228-29.

Lancaster, Roy 1981. *Plant Hunting in Nepal.* Vikas Publishing House, New Delhi.

Law, Y.W. 1947. Gymnosperms of Eastern China. *Bot. Bull. Academia, Sinica,* 1; 141-71.

Lawrence, G.H.M. 1960. *Taxonomy of Vascular Plants.* ed. 5. Macmillan Coy New York.

Lee, Shun Ching 1935. *Forest Botany of China.* Shanghai.

Li, 1952. J. Arn. Arbor. 33(2): 192-8.

Long, D.G. 1979. The Bhutanese Itineraries of William Griffith and R.E. Cooper. *Notes R.B.G. Edinburgh* 37: 355-68.

Macdonald 1982. *The Macdonald Encyclopedia of Trees.* Macdonald & Co. London & Sydney.

Maheshwari, P. & Biswas, C. 1970. *Cedrus,* Bot. Monogr. No. 5: 1-115, CSIR, New Delhi.

Mayr, Heinrich. Wald und Parkbaume Fuer Europa 105, 174.

Mehra, P.N. & Jain, K.K. 1976. Abies and Juniperus complexes in the E. Himalaya with observations on Larix griffithii Hook.f. and Tsuga dumosa Eichler. Punjab University, Chandigarh (India).

Melchior, H. & Werdermann, E. 1954. In Englers Syllabus der Pflanzenfamilien ed. 12.1.34.

Merrill, E.D. 1941. The Upper Burma Plants collected by F. Kingdon-Ward on the Vernay-Cutting Expedition. Brittonia, 4: 26.

Moule, A.C. 1944. The name Ginkgo. *J. Roy. Hort. Soc.* 69: 166.

Nair, N.C. 1977. *Flora of Bashahr Himalayas.* International Bioscience Publishers, Hissar.

Narayanamurti. D *et al* 1941. *Indian For. Leafl.,* No. 6.

Orr, M.Y. 1933. Plants Chinensis Forrestianae. Coniferae *Notes R. Bot. Gdn.* Nov. XVIII (LXXXVII) 131.

______, Ibid. Coniferae Notes R.B.G. Edinb. Nov. 142-43.

Page, C.N. & Rushforth, K.D. 1980. Picea farreri, a new temperate conifer from Upper Burma. *Notes R.B.G. Edinb.* 38(1): 129-136.

Pant, D.D. 1957. The classification of gymnospermous plants. *Palaeobotanist,* 6: 65-70.

______, 1973. *Cycas and the Cycadales.* Central Book Depot, Allahabad.

______ & Nautial, D.D. 1963. Cuticular and epidermal studies of some modern Cycadean leaves, etc. *Senck. Biol.,* 44(4): 257-347.

Parker, R.N. 1927. The Himalayan Silver Firs and Spruces. *Indian For.* 52(12): 683-93.

______, 1940. Abies spectabilis Spach & A. pindrow Spach. Ibid. LXVI(1): 1-3.

Pilger, R. 1916. Kritische Ubersicht uber die neuere Literatur betraffend die Familie der *Taxaceae.* Bot. Jb. 54: 1-43.

______, 1926. Gymnospermae. In Engler & Prantl, Die naturlichen Pflanzenfamilien ed. 2. 13: 1-447. Berlin.

Pulle, A. 1938. The classification of the spermatophytes. *Chron. Bot.* 4: 109-13.

Pulle, A.A. 1940-46. Over de *Ginkgo* alias *Ginkgo.* Jaarbock Nederl Dendrd. Ver., pp. 25-35.

Polunin, O. & Stainton, A. 1984. *Flowers of the Himalaya.* Oxford University Press Delhi, Bombay, Calcutta, Madras.

Raizada, M.B. & Sahni, K.C. 1960. *Living Indian Gymnosperms.* pt. 1. (Cycadales, Ginkgoales, Coniferales. *Indian For. Rec.* 5(2): I-IV, 73-150.

Rao, A.R. 1953. *The Palaeobotanist,* 2: 25-23.

Rehder, A. 1940. 2nd edn. *Manual* of *Cultivated Trees & Shrubs Hardy in N. America,* Macmillan Publishing Co. Inc, New York.

Rehman, M.A. 1945. Indian Woods for Pencil Making. *Indian For. Leaflet* No. 66.

Rheede, H.V. 1688. *Hortus Malabaricus.*

Roxburgh, W. 1832. *Flora Indica.* Calcutta.

Sahni, B. 1920a. On the structure and affinities of *Acmopyle Pancheri Pilger. Phil. Trans. Roy. Soc.* London 210(B): 253-310.

Sahni, B. 1920b. On Certain archaic features in the seed of *Taxus baccata* with remarks on the antiquity of the Taxineae. *Ann. Bot.* 34: 117-133.

Sahni, B. 1928. Revision of Indian Fossil Plants Pt. I. *Coniferales* (a Impressions and Incrustations) *Mem. geol. Sury. Ind. Palaeont. Indica* (N.S.) 11: 1-49.

______, 1931. Revision of Indian Fossil Plants Pt II *Coniferales* (b. Petrifications *Mem. geol. Surv. Ind. Palaeont. Indica* (N.S.) 11: 51-120.

______, 1948. The *Pentoxyleae* : A new group of Jurassic Gymnosperms from the Rajmahal hills of India. *Bot. Gaz.* 110: 47-80.

Sahni, K.C. 1953. Botanical Exploration in the Great Nicobar Island. *Indian For.* 79(1): 3-16.

______, 1962. A conifer new to the Flora of India. *Indian For.* 88(10): 748-49.

______, 1969. Contribution to the Flora of Kameng and Subansiri Dists, NEFA. *Indian For.* 95(5): 330-52.

Sahni, K.C. 1986. Gymnosperms in Encyclopedia of Indian Natural History pp. 129-134. General Edit. R.E. Hawkins, Oxford University Press, Delhi, Bombay.

______, 1970. Protection of Rare and Endangered Plants in the Indian Flora. *Proc IUCN* 11th *Tech. Meeting Morges Switzerland.* 11: 95-102.

______, 1977 April, 1973 Oct. Tour Reports, Kameng District, Arunachal Pradesh (unpublished).

______, 1989. A new variety of Blue Pine from the E. Himalaya *Indian J. For.* 12(1) 40-42.

Santapau, H. & Henry, A.N. 1973. *A. Dictionary of the Flowering Plants in India.* CSIR, New Delhi.

Schuster, J. 1932. Pflanzenreich *Cycadaceae* Leipzing.

Schwarz, O. 1938. Notiz Zur Nomenklatur von Pinus excelsa Wall. Fedde Repert XLIV: 160.

Sporne, K.R. 1965. *The Morphology of Gymnosperms.* Hutchinson & Co. (Publishers) London.

Sprague, T.A. (1933. Abies pindrow Royle Kew Bull., 414.

Srivastava, B.P. 1945. Silicified plant remains from the Rajmahal series of India *Proc.* Nat. *Acad. Sci. India,* **15:** 185-211.

Stearn, W.T. 1938. *Larix Griffithii* or *Larix Griffithiana? 150th Anniversary, Vol. R. Bot. Gdn. Calcutta* 115.

Takhtajan, A.L. 1953. *Phylogenetic Principles of the System of Higher Plants. Bot. Rev.,* **19**: 1-45.

Thierot, J.W. 1980. *The New Encyclopaedia Britannica* **8:** 518-24. 15ed. Helen Hemingway Barton, Publisher, London.

Thommen, E. 1949. Neuse Zur Schreibung des namens Ginkgo. ver. Nat. Ges Basel 66: 77-103.

Thothathri, K. *et al.* 1973. Botanical Results of the Joint Scientific Expedition to the Great Nicobar Island. *Bull. Bot. Surv. India* 15(3 & 4): 235-265.

Troll, Carl 1939. Das Pflanzenkleid des Nanga Parbat begleitworte Zur Vegetationskarte der Nanga Parbat Gruppe Sonder druck aus Wissenchaftliche Veroffentlichungan des Deutschen Kunde Zu Leipzig N.F. 7.

Troup, R.S. 1921. *Silviculture of Indian Trees* III, Oxford University Press, Oxford.

Tudge, C. 1988. *The Encyclopedia of the Environment.* Christopher Helm London.

Van Tieghem. Ph 1898. *Elements de Botanique.* Ed. 3, Paris.

Vasil, V. 1959. Morphology and embryology of *Gnetum ula* Brongn. *Phytomorphology* 9: 167-214.

Vigneg, M. Th. et Gaussen H. 1929. Revision du genre *Abies.* Travaux du Laboratorie Forestier de Toulouse.

Wealth India - Raw Materials (1948-76. Pts 1-XI, CSIR, New Delhi.

Wan-Chun, Cheng et Li-kuo Fu, Edit. 1978. Flora Reipublicae Popularis Sinicae, Tomus 7, Gymnospermae, Science Publications, Beijing, 544 pp.

Widder, F. 1948. Die Reichtschreibung des Namens *Ginkgo. Phyton* 1: 47-52.

Willis, J.C., 8th edn. by H.K. Airy Shaw 1973. *A Dictionary of Flowering Plants and Ferns.* Cambridge University Press, Cambridge, New York.

______, & Gardiner, J.S. 1901. The botany of the Maldive Isls. Ann Roy. Bot. Gard. Peradeniya, Ceylon. Vol. 1. pt. II.

Wilson, E.H. 1926. The Taxads and Conifer of Yunnan. *J. Arnold Arb.,* VII(1); 37-68.

INDEX OF LATIN NAMES

INDEX TO ENGLISH NAMES

INDEX TO VERNACULAR NAMES

PLATE I

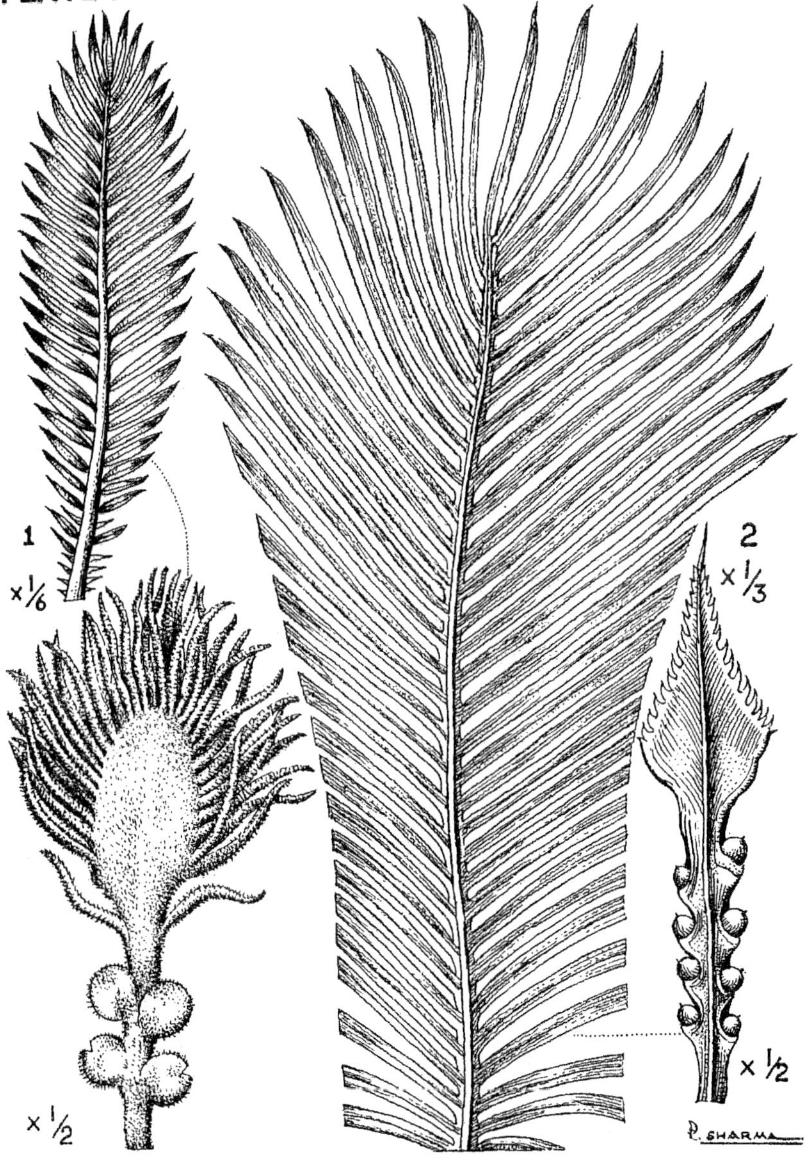

1. Cycas revoluta. 2. C. circinnalis (Leaves and megasporophylls).
FAMILY 1 CYCADCEAE

PLATE 2

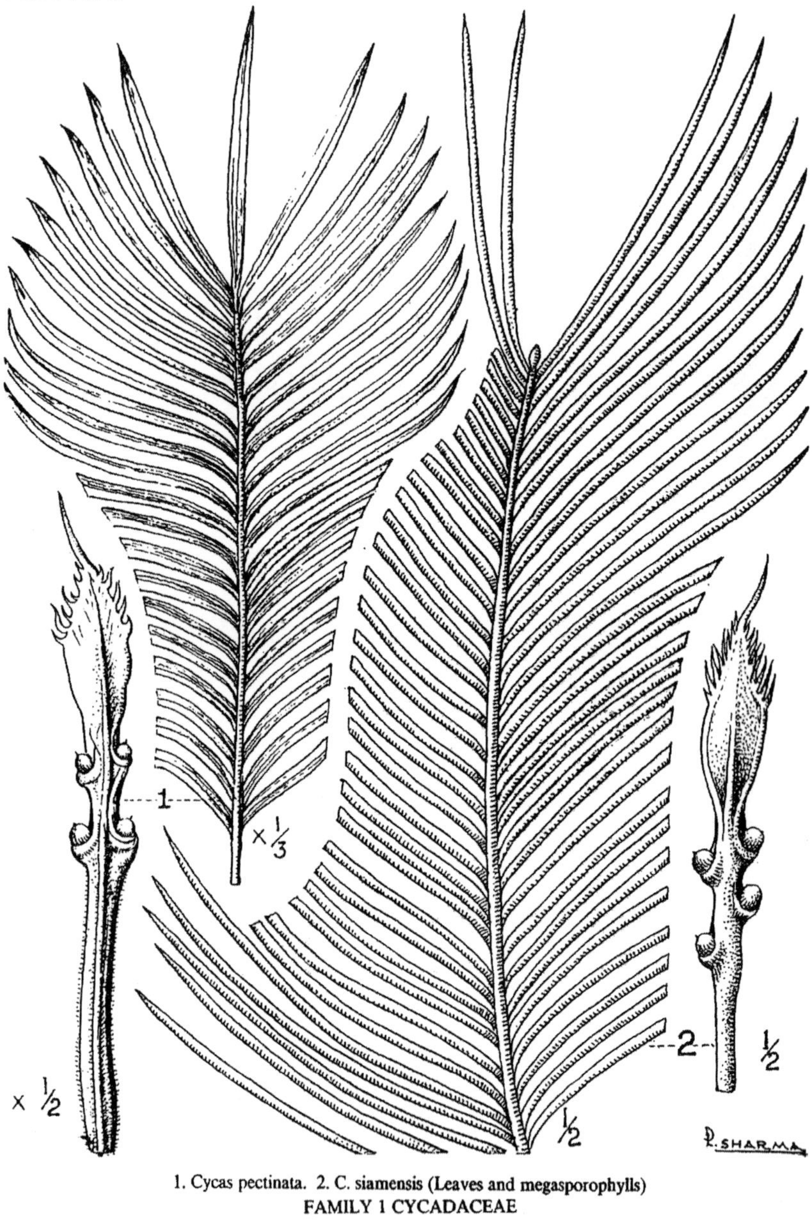

1. Cycas pectinata. 2. C. siamensis (Leaves and megasporophylls)
FAMILY 1 CYCADACEAE

PLATE 3

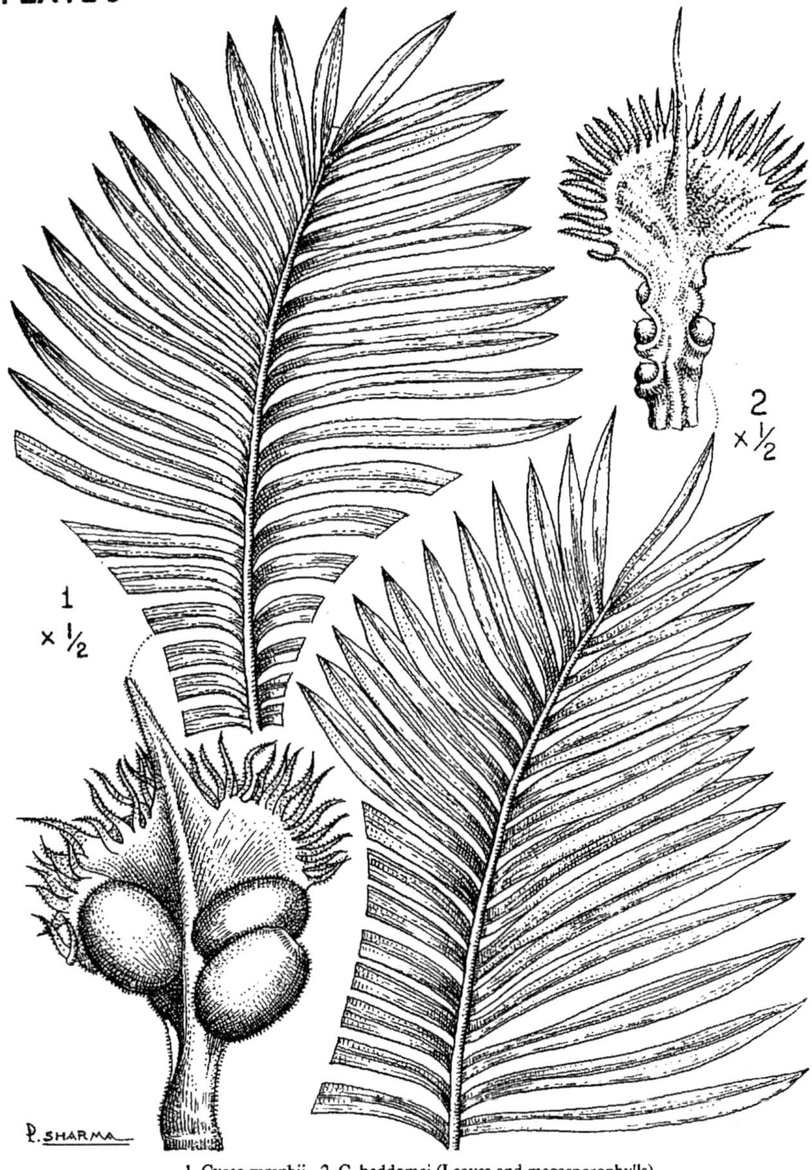

1. Cycas rumphii. 2. C. beddomei (Leaves and megasporophylls).
FAMILY 1 CYCADACEAE

PLATE 4

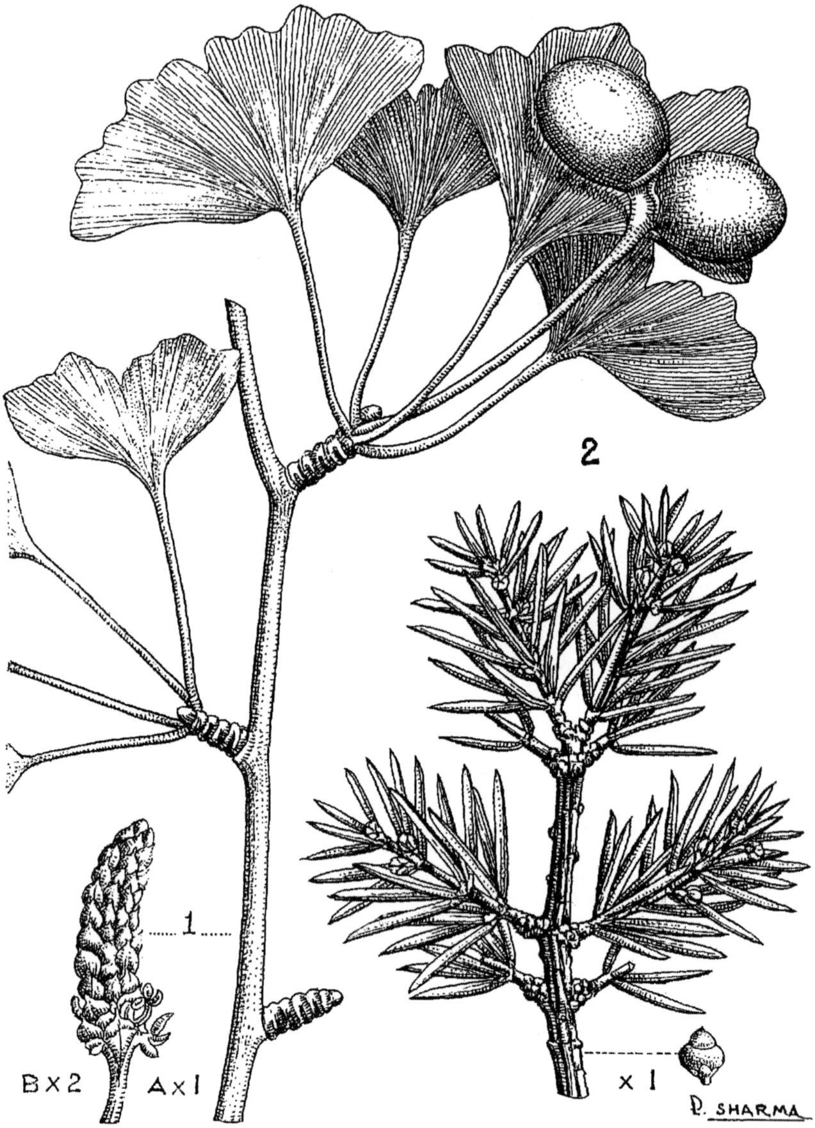

1. Ginkgo biloba. A. Shoot with leaves and a pair of ovules. B. Male strtobilus showing sporophylls each bearing two sporangia

FAMILY 2 GINKGOACEAE

2. *Taxus baccata*, shoot with male flowers; seed.

FAMILY 3 TAXACEAE

PLATE 5

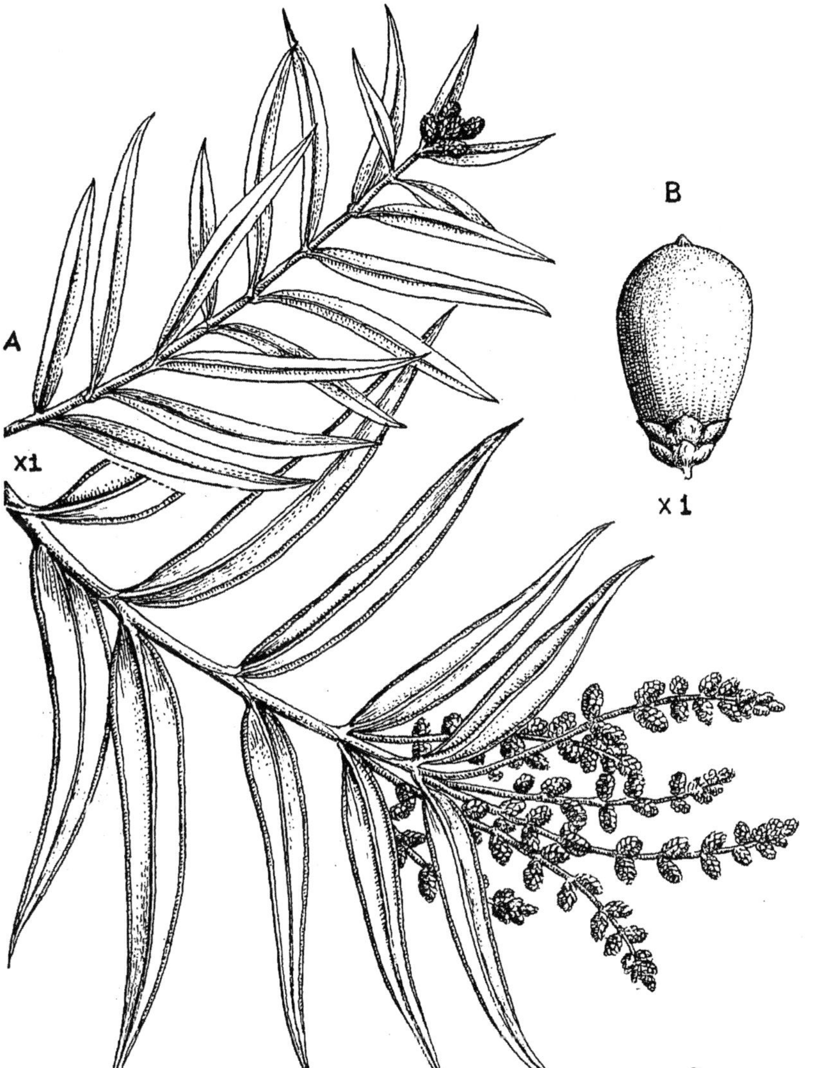

Amentotaxus assamica : Branchlet showing distichous leaves and terminal immature, staminate racemes. Leaves linear lanceolate commonly, 7-11.5 × 0.38 cm when mature. (After Ferguson K.B. 40 (1); 17, 1985, *Kingdon Ward* 8026). B. after *Haridasan* 1443 Fr. 2-3.5 × 1.5-2.5 cm shortly beaked.

FAMILY 3 TAXACEAE

PLATE 6

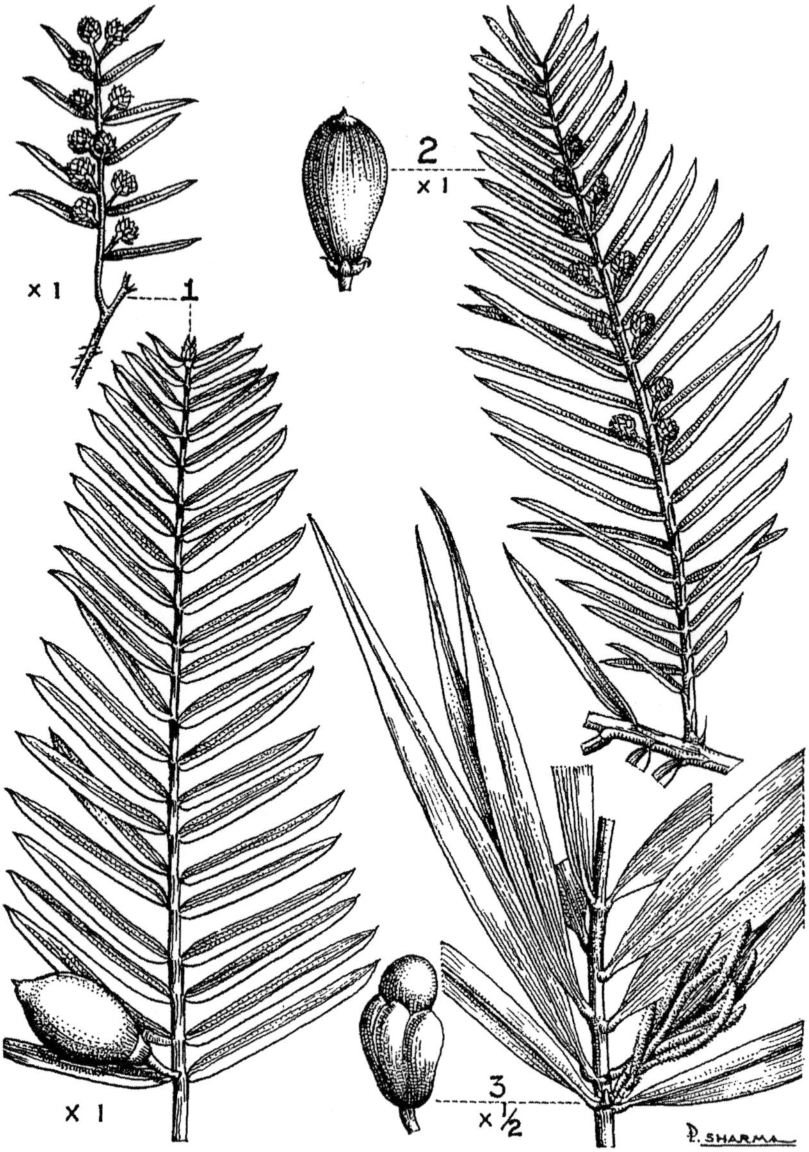

1. *Cephalotaxus griffithii*, shoot with opposite leaves, male strobli in globose heads, seed with an apical tip. 2. *C. manni*, folige, male strobili, female cone.

FAMILY 4 CEPHALOTAXACEAE

3. *Podocarpus neriifolius* Leaves oleander-like, male strobilus in spicate clusters, seed on a fleshy receptacle.

FAMILY 5 PODOCARPACEAE

PLATE 7

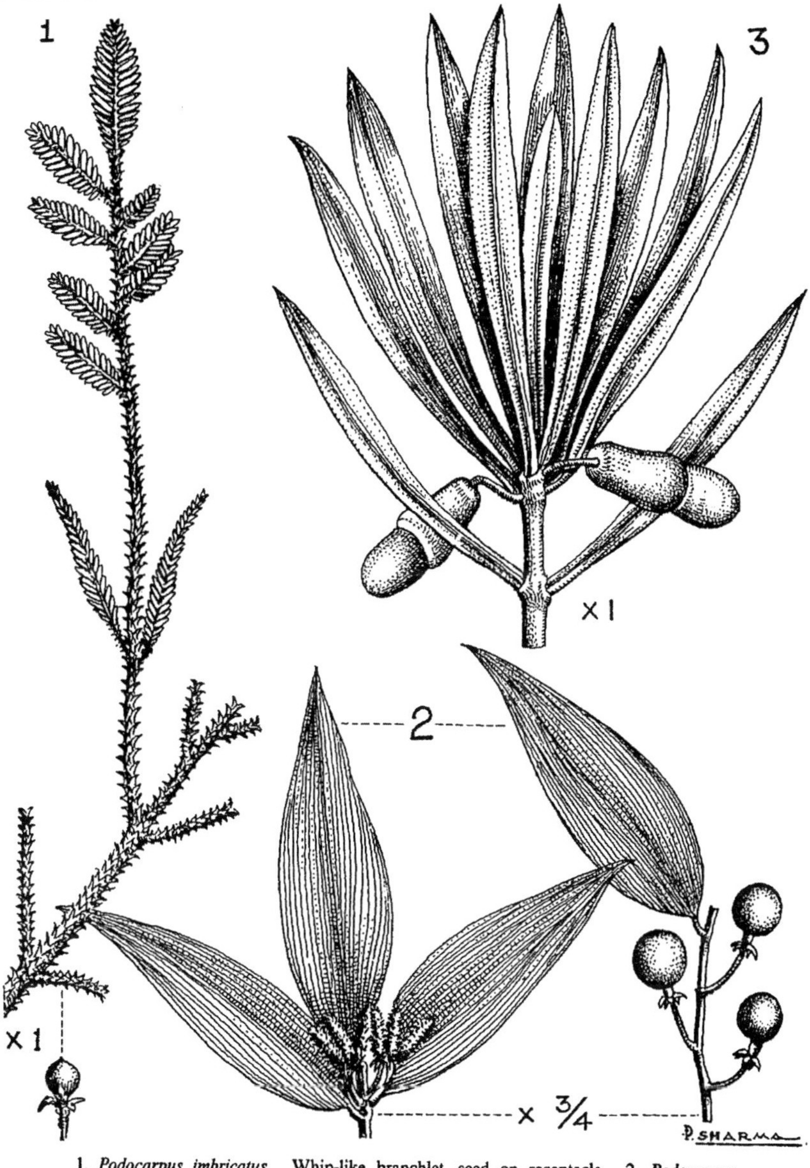

1. *Podocarpus imbricatus.* Whip-like branchlet, seed on receptacle. 2. *Podocarpus wallichianus.* Foliage, male strobili in spikes, 3 seeds. 3. *Podocarpus macrophyllus.* Foliage, seeds borne on receptacles.

FAMILY 5 PODOCARPACEAE

PLATE 8

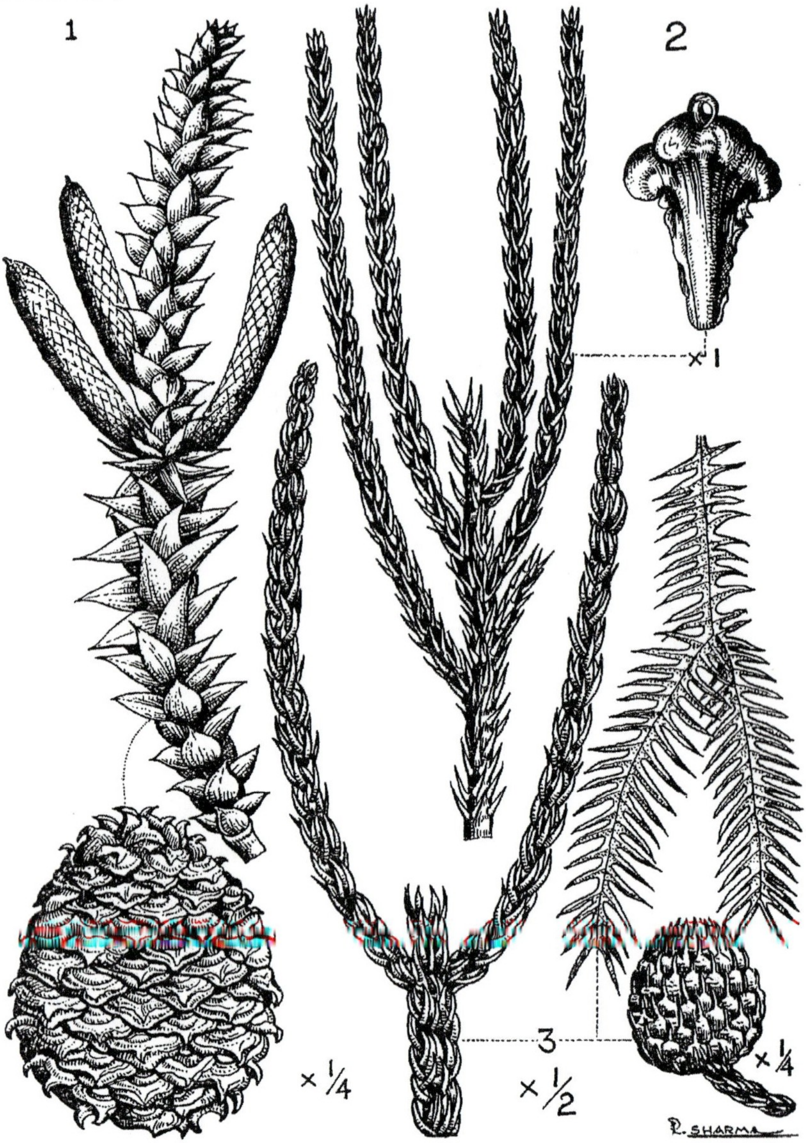

1. *Araucaria bidwillii.* Shoot with foliage and male catkins; female cone. 2. *A. columnaris.* Shoot with soft foliage, cone scale. 3. *A. cunninghamii.* Leaves of two kinds; on young trees spirally arranged on old trees, coning branches, shorter, crowded and sharp pointed.

FAMILY 6 ARAUCARIACEAE

PLATE 9

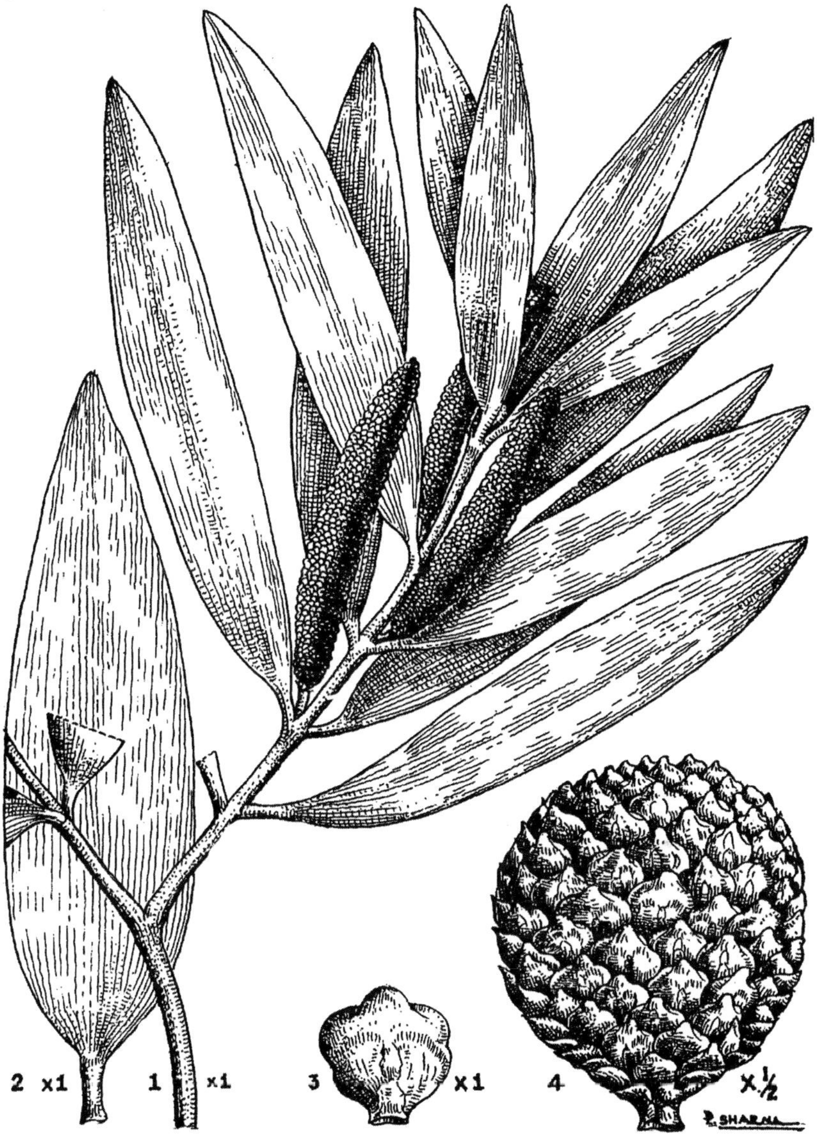

1. *Agathis robusta;* Male cones, foliage. 2. Leaf. 3. ♀ cone scale. 4. ♀ cone × ½ (All figures from *Vaid* s.n. and *Pant* 6302 in DD)

FAMILY 6 ARAUCARIACEAE

PLATE 10

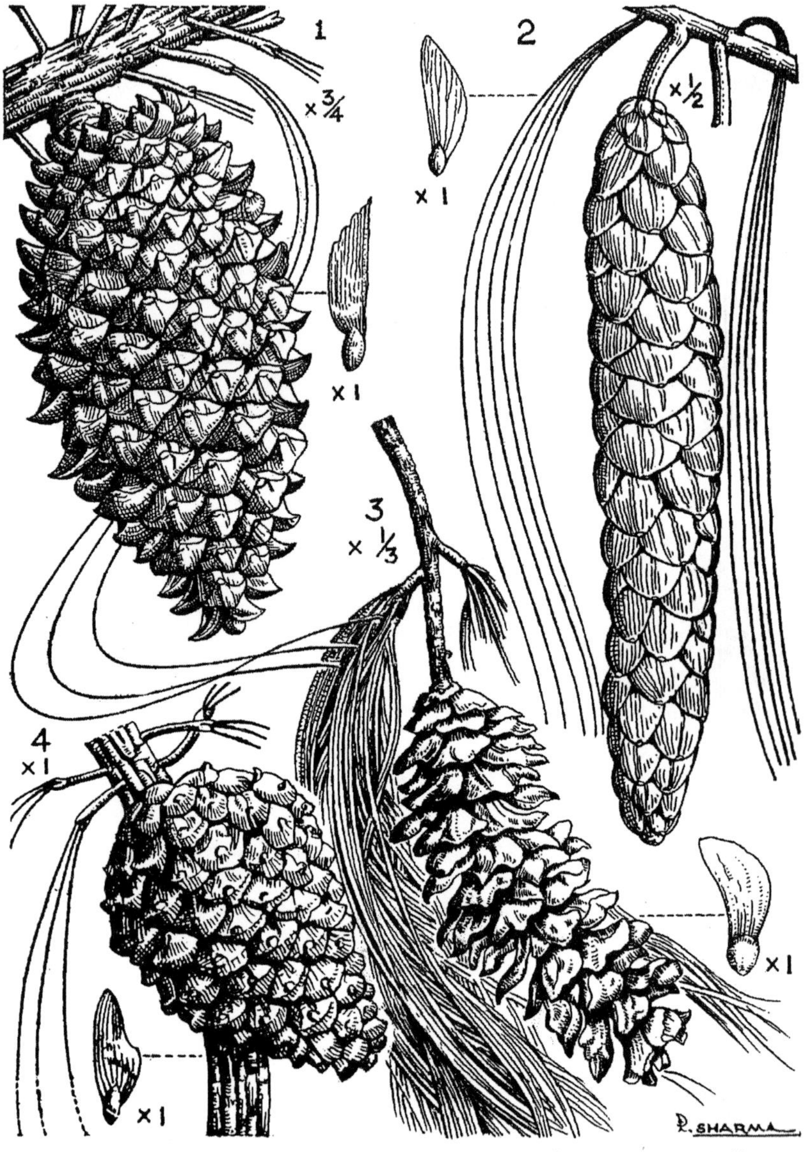

1. *Pinus roxburghii:* Needles 3, ♀ cone with thick woody, pyramid shaped scales, seed with long wing. 2. *Pinus wallichiana :* Needles 5, ♀ cone banana-shaped, scales adhering like scales of fish. Cone ripening not yet open. October. 3. *P. bhutanica :* Needles long pendulous mature open cone, seed with wing, seed 6-8 mm, wing 2 cm. 4. *P. kesiya :* Needles 3, ♀ cone seed 8 mm, wing 4 times the length of seed.

FAMILY 7 PINACEAE

PLATE 11

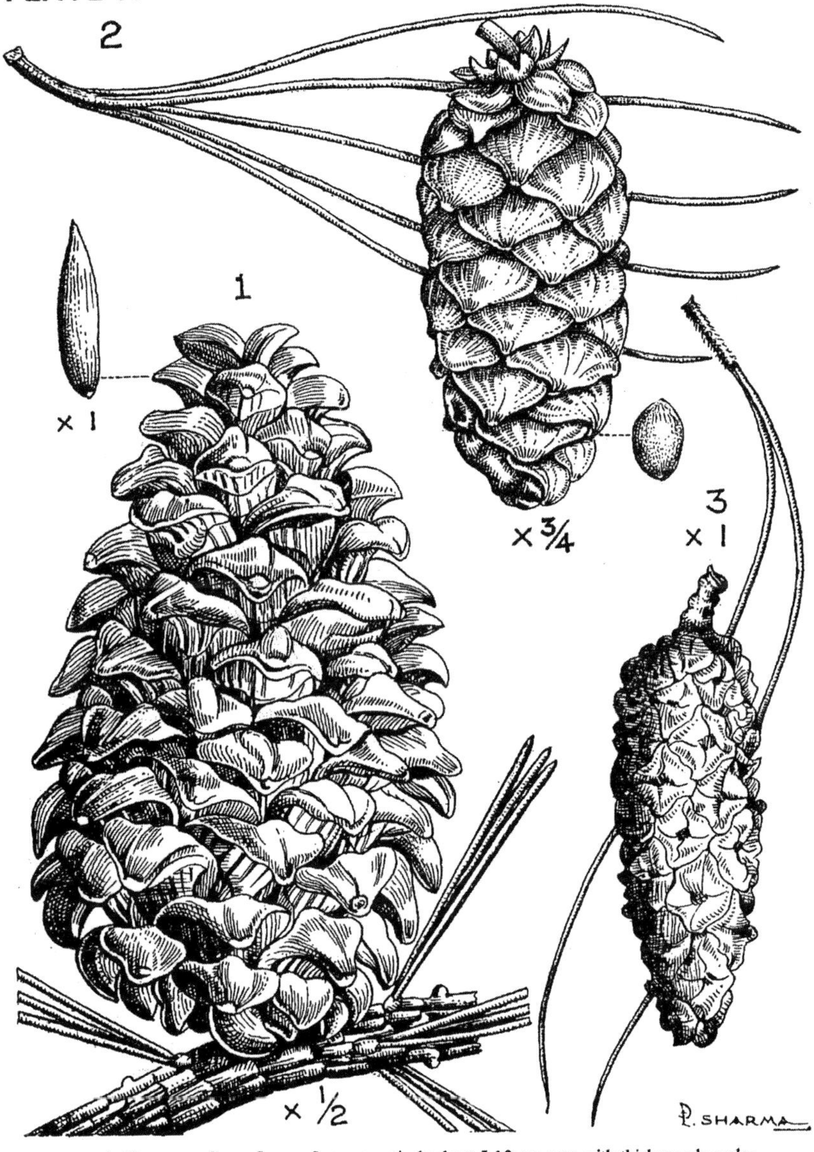

1. *Pinus gerardiana :* Leaves 8, comparatively short 5-10 cm, cone with thick woody scales with a recurved back. Seed cylindric. 2. *P. armandii* : Leaves 5, cone barrel-shaped 10-18 cm long, seed. 3. *P. merkusii :* Leaves 2, cone cylindric conical 5-7.5 cm long

FAMILY 7 PINACEAE

PLATE 12

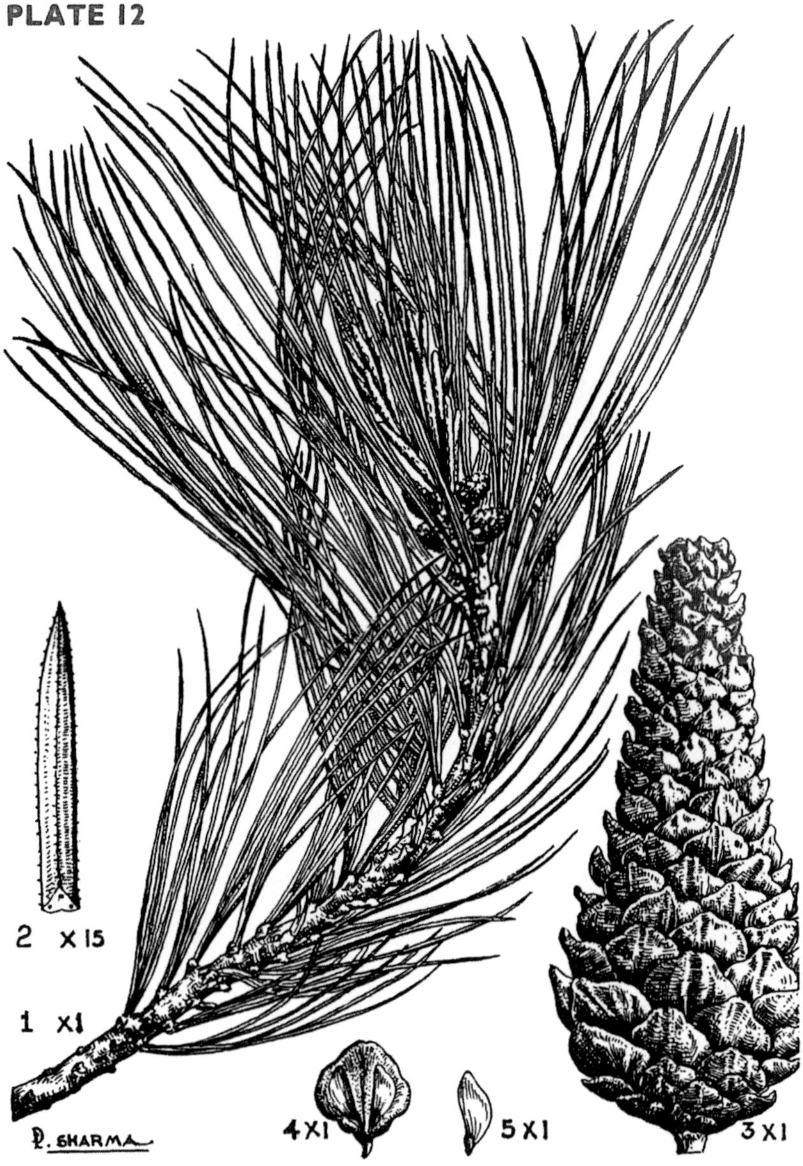

Pinus wallichiana A.B. Jacks., var. *parva* K.C. Sahni. 1. Shoot with foliage and male strobili. 2. Needle cut in the middle showing dorsal resin canals and one ventral. 3. ♀ cone 10 cm long. 4. Cone scale with 2, winged seeds. 5. Seed 3 mm wing 10 mm.
FAMILY 7 PINACEAE

PLATE 13

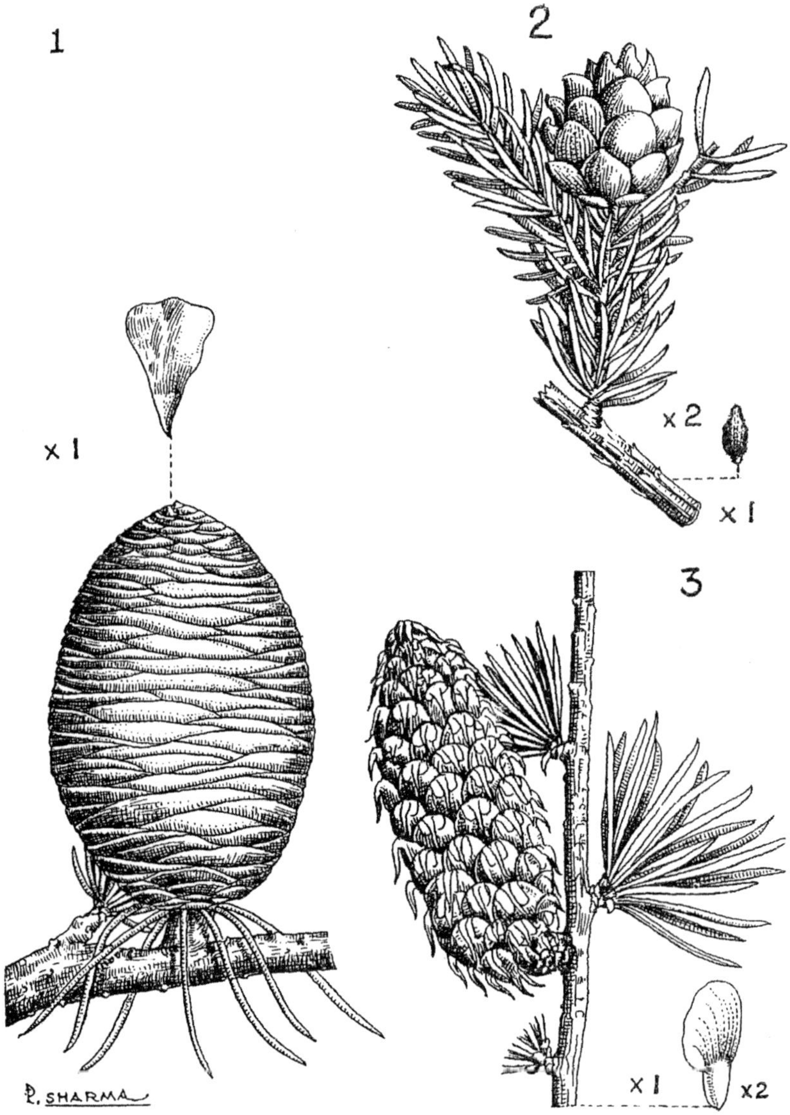

1. *Cedrus deodara* Leaves in tufts of 10-15, 2.5-3.8 cm. Cone barrel-shaped *c* 7.6 × 5.8 cm. Seed triangular with a broad wing seed and wing 2.5-3.2 cm. 2. *Tsuga dumosa* : Shoot with foliage, leaves 1.27-2.5 cm cone ovoid 2.5 cm, seed 0.88 cm, wing oblong. 3. *Larix griffithiana* : Needles in tufts of 10-19, cone 5-10 × 2.5-3.2 cm. Seed 0. 4 cm, wing 1.2 cm.

FAMILY 7 PINACEAE

PLATE 14

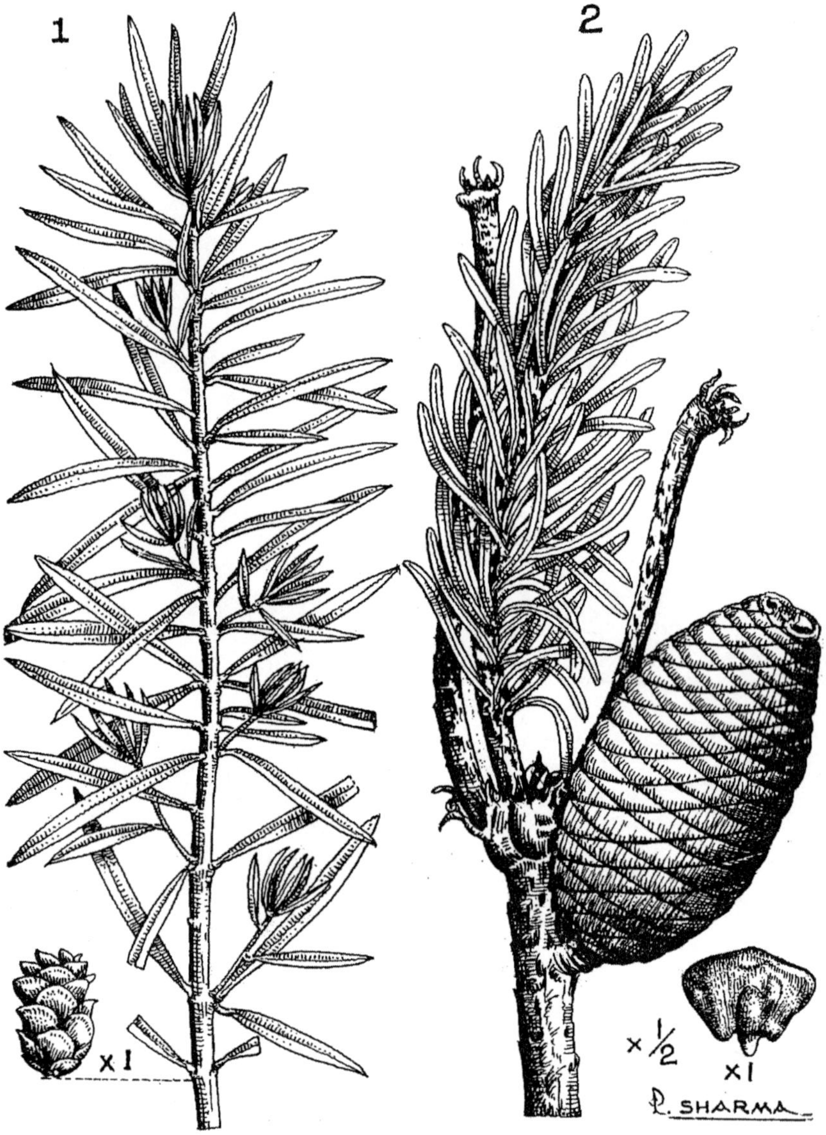

1. *Tsuga yunnanensis :* Shoot and female cone with roundish scales. 2. *Abies densa :* Shoot with crowded leaves, female cone, fan-shaped cone scale with seed.

FAMILY 7 PINACEAE

PLATE 15

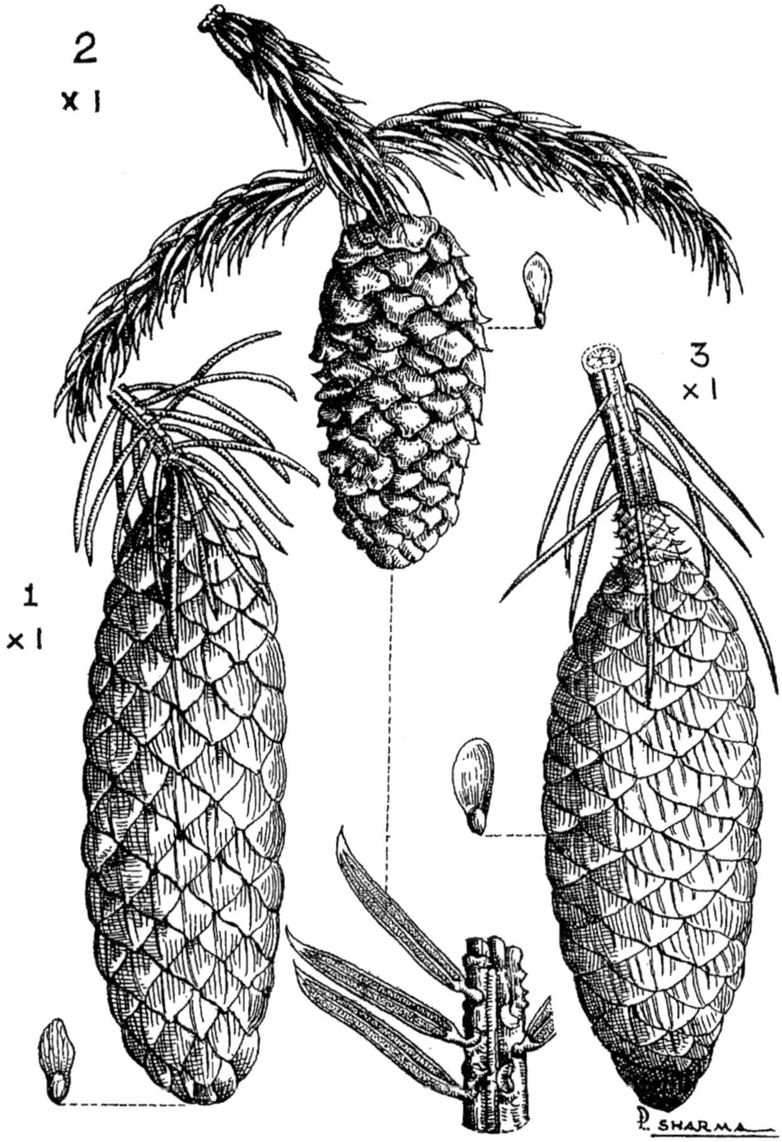

1. *Picea spinulosa :* Needles 1.9-3.2 cm, flatened sharply pointed. Cone cylindric 10 × 3 cm. Seed with wing 1.27 cm. 2. *Picea brachytyla :* Leaves crowded sharply pointed. Cone cylindric oblong 6.3 cm-8.8 cm. Scales broad and rounded at the summit, apex wavy. 3. *Picea smithiana :* Needles 2.5-3.8 cm long, 4 sided sharply pointed. Cone 10 × 3.5 cm seed 1.27 cm including the wing.

FAMILY 7 PINACEAE

PLATE 16

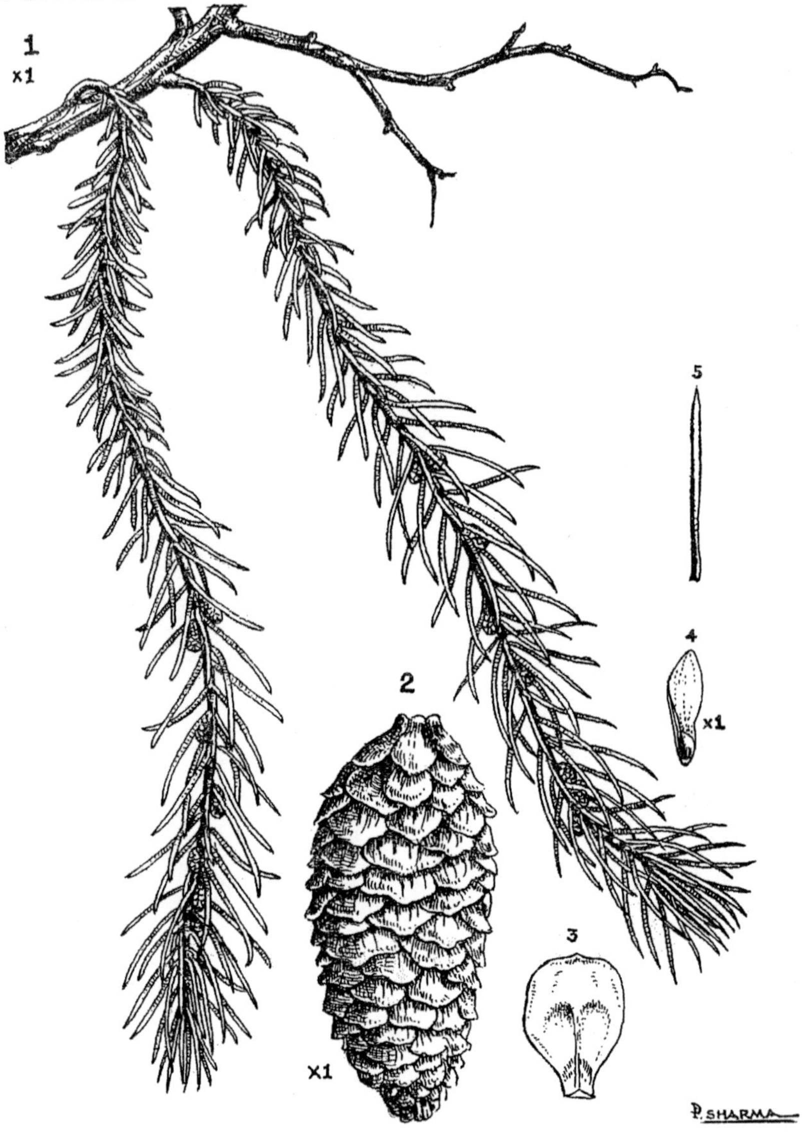

1. *Picea farreri* : A spray showing foliage. 2. ♀ cone, ellipsoid cylindric 7 × 3 cm. 3. cone scale. 4. Seed with wing. 5. Leaf (enlarged).

FAMILY 7 PINACEAE

PLATE 17

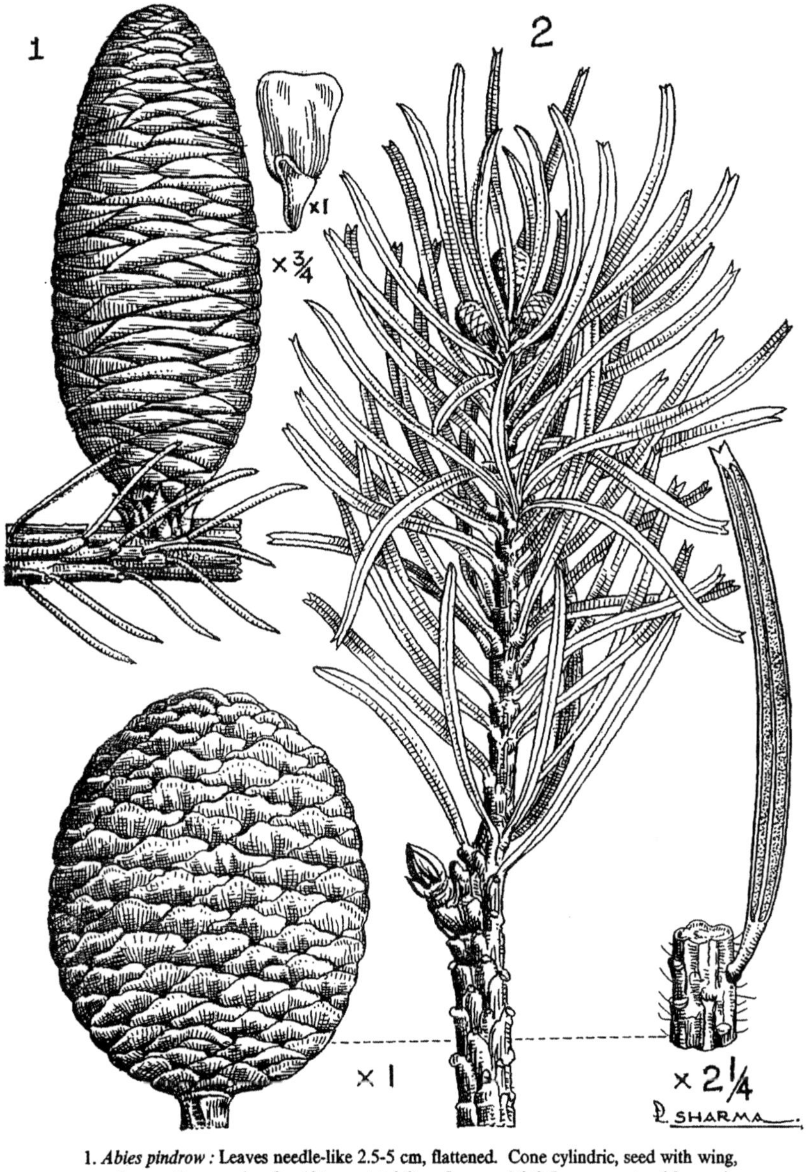

1. *Abies pindrow :* Leaves needle-like 2.5-5 cm, flattened. Cone cylindric, seed with wing, longer than seed. 2. *Abies spectabilis :* Leaves 1.3-2.5 cm, up to 6.2 cm in E. Himalaya. Cones shorter, thicker than *A. pindrow c* 7 cm long.

FAMILY 7 PINACEAE

PLATE 18

1

× 1

1. ***Abies delavayi*** **:** Leaves 1.8-3.1 cm. ♀ cone *c* 6.27 cm, broadly cylindrical or barrel-shaped.

FAMILY 7 PINACEAE

PLATE 19

Larix himalaica : Spray showing foliage, ♀ cones. Winged seed.

FAMILY 7 PINACEAE

PLATE 20

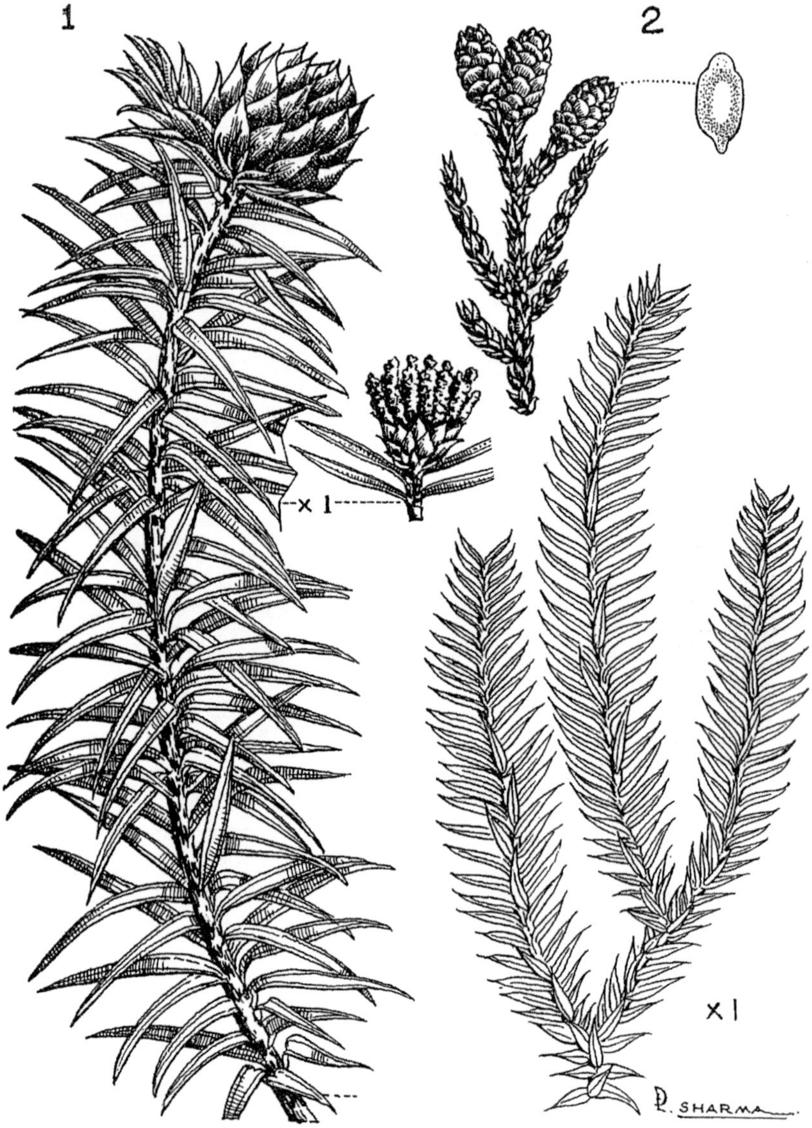

1. *Cunninghamia lanceolata* Shoot with foliage, female cone. A cluster of male cones.
2. *Taiwania cryptomerioides :* Shoots with 2 types of leaves of young trees narrowly linear (bottom right), adult are thick triangular. ♀ cone cylindric, *c* 1.27 cm long

FAMILY 8 TAXODIACEAE

PLATE 21

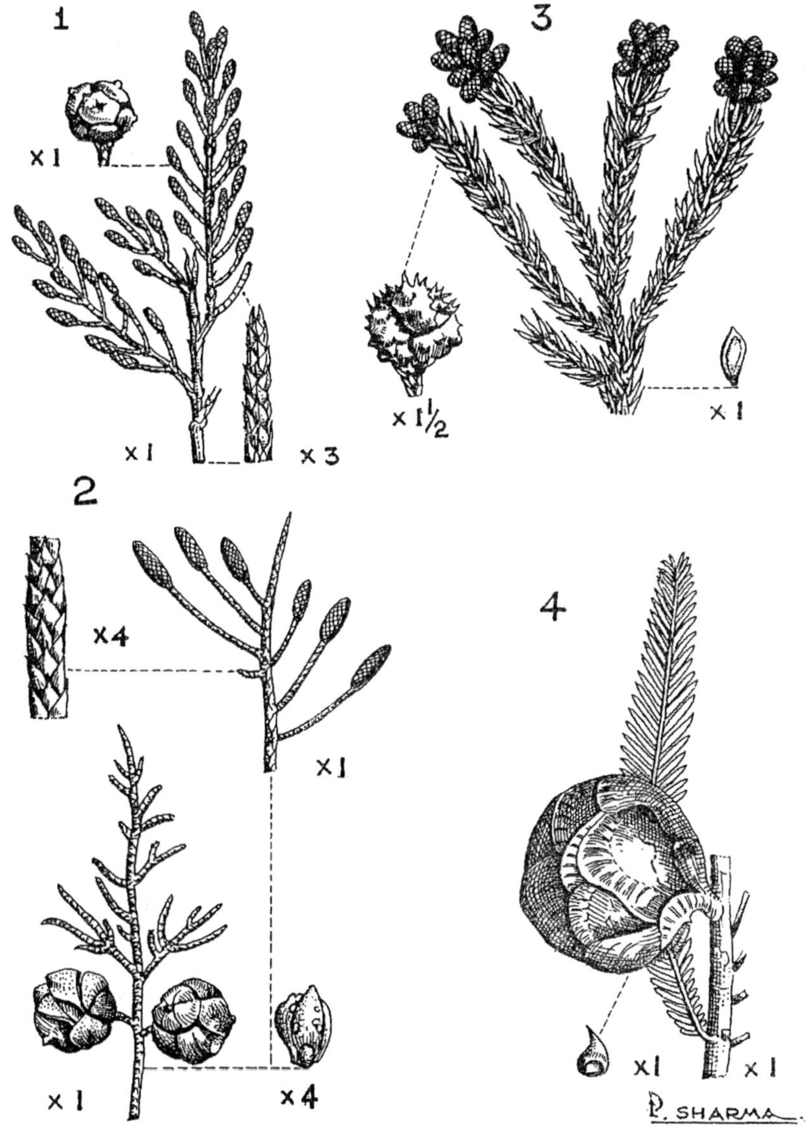

1. *Cupressus funebris :* Shoot with male strobili. Branchlet magnified × 3 showing scale leaves. ♀ cone. 2. *Cupressus torulosa :* Two sprays showing cylindrical male strobili, two female cones seed magnified × 4. Seed marked with resin glands.

FAMILY 9 CUPRESSACEAE

3. *Cryptomeria japonica :* Branchlet with awl-shaped, apirally arranged needles. Terminal clusters of male strobili. Female woody cone (left). 4. *Taxodium distichum :* Branchlet with leaves and ♀ cone.

FAMILY 8 TAXODIACEAE,

PLATE 22

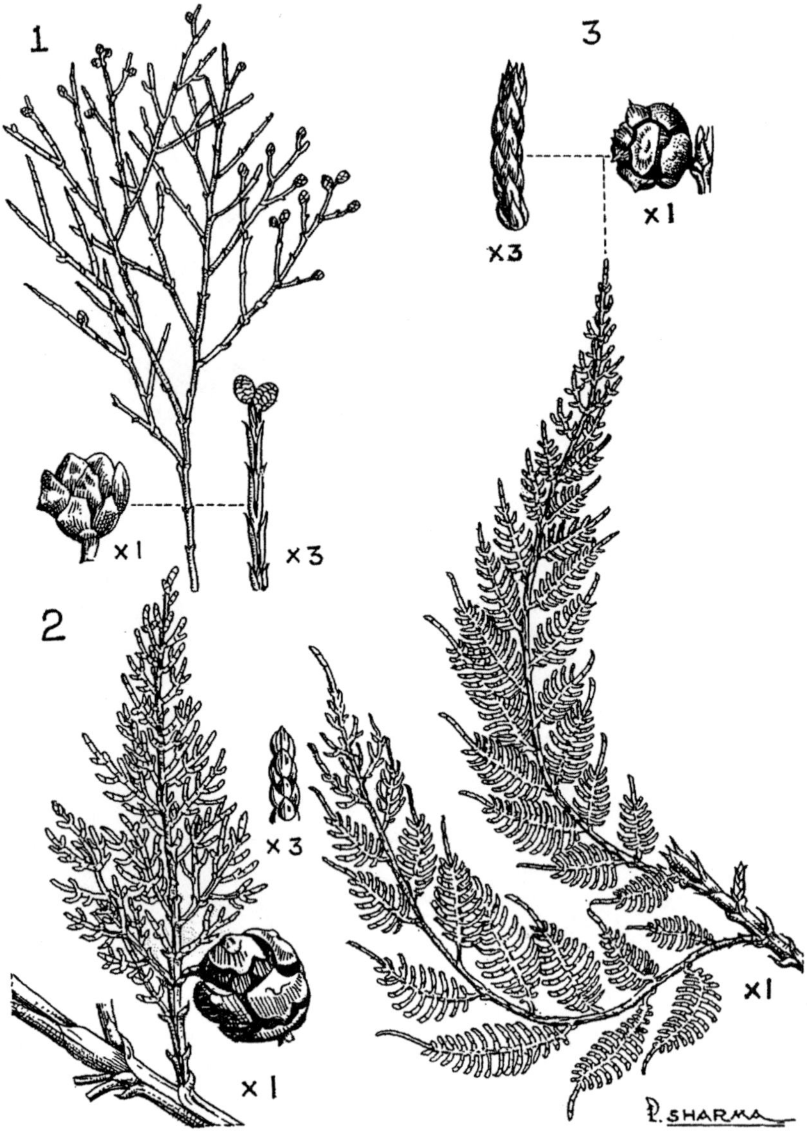

1. *Callitris rhomboidea* : Branchlet with male strobilus. Femle cone. 2. *Cupressus sempervirens* : Branchlet with ripe cone. 3. *Cupressus cashmeriana* Fan shaped branchlets. ♀ cone.

FAMILY 9 CUPRESSACEAE

PLATE 23

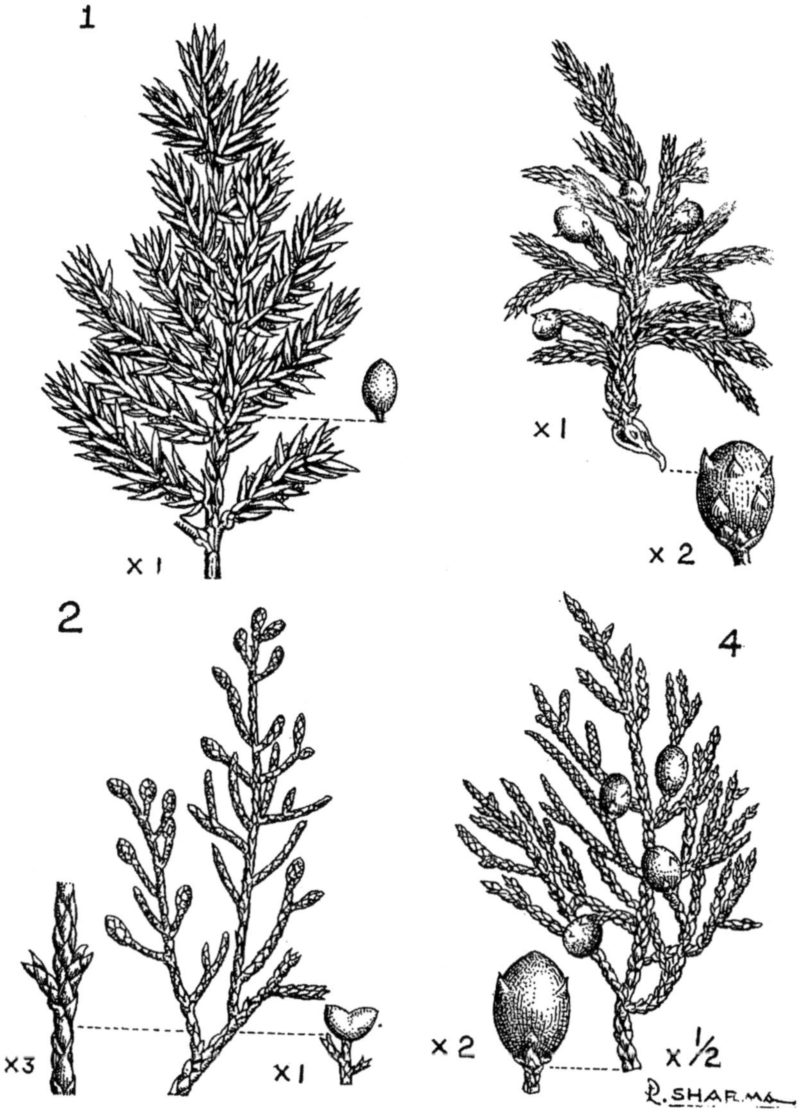

1. *Juniperus communis* : Branchlet with sharply pointed awl-shaped leaves, fruit ovoid. 2. *Juniperus polycarpos* : Branchlet with male catkins, fruit. 3. *Juniperus recurva* : Branchlet with fruits. 4. *Juniperus indica* : Branchlet with fruits.

FAMILY 9 CUPRESSACEAE

PLATE 24

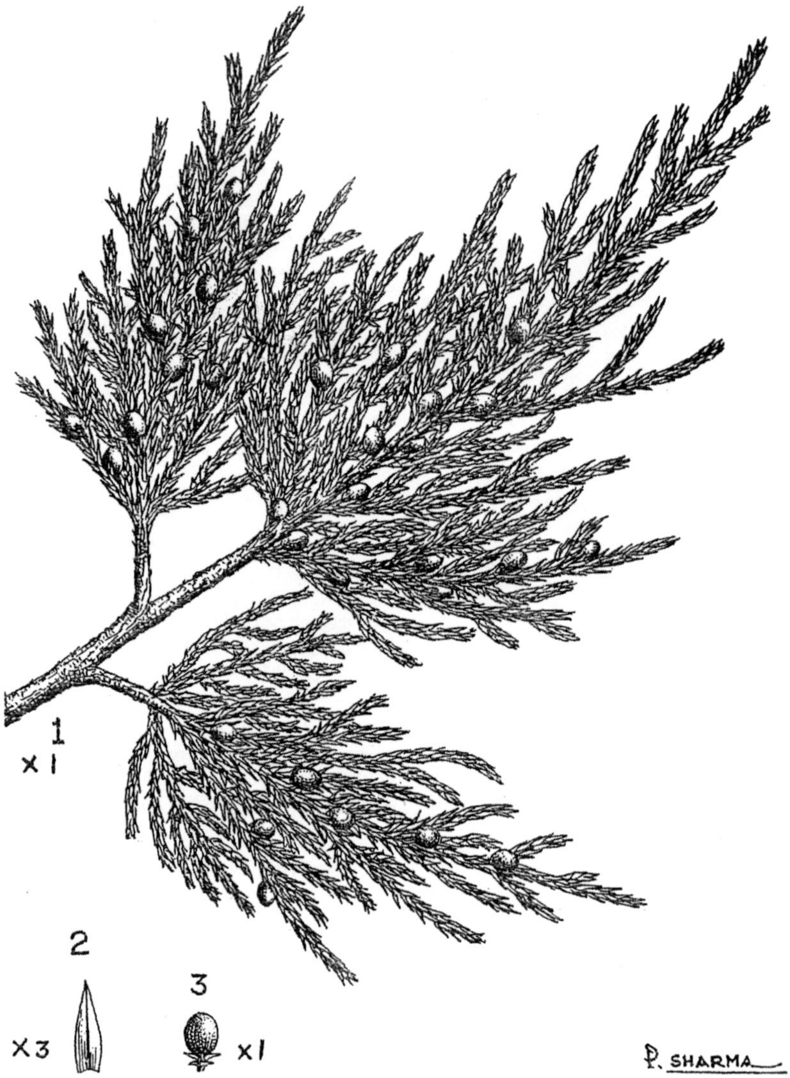

Juniperus squamata 1. Branchlet with awl-shaped overlapping leaves and many fruits. 2. A magnified leaf. 3. An ellipsoid fruit *c* 0.6 cm long.

FAMILY 9 CUPRESSACEAE

PLATE 25

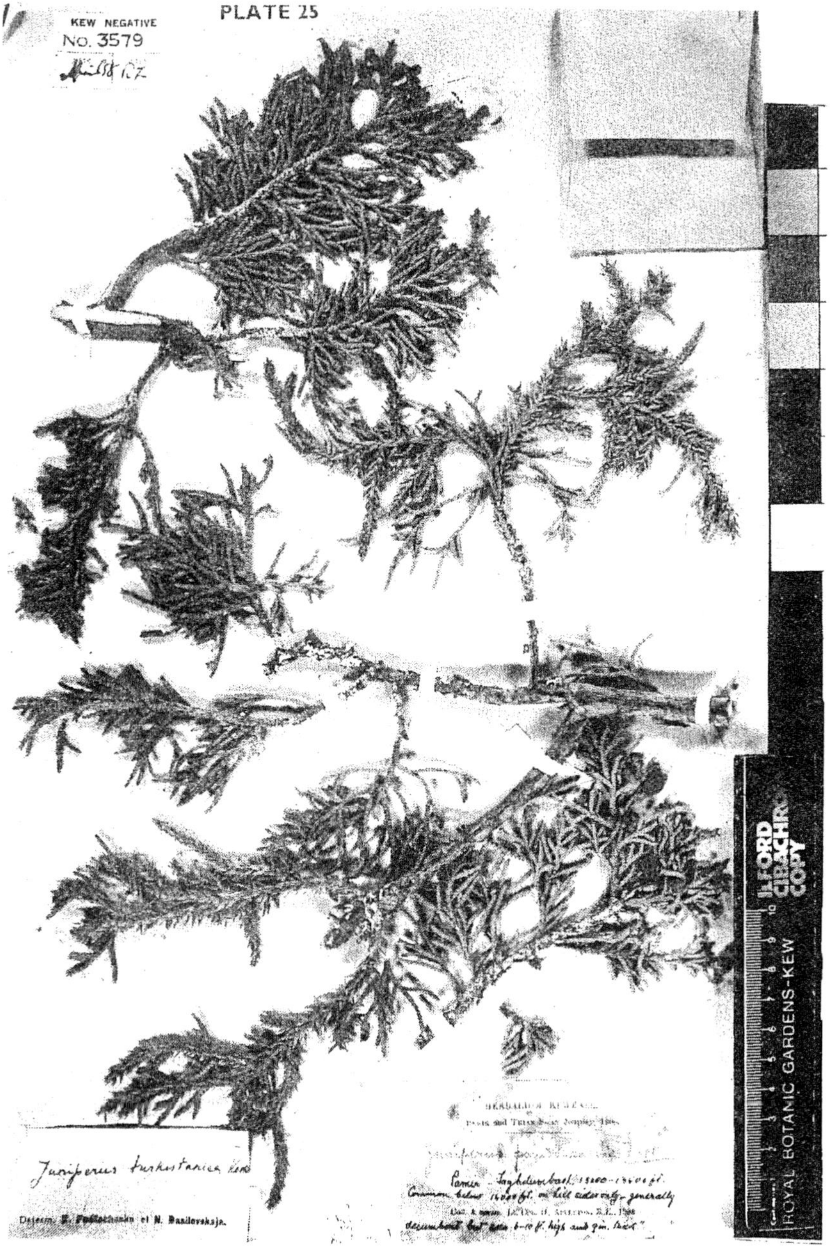

Juniperus turkestanica Komorov. Pamirs C. Asia. Coll. Lt. Col. H. Appleton in 1908. 13000-13500 ft. Courtesy, Herbarium Kew.

FAMILY 9 CUPRESSACEAE

PLATE 26

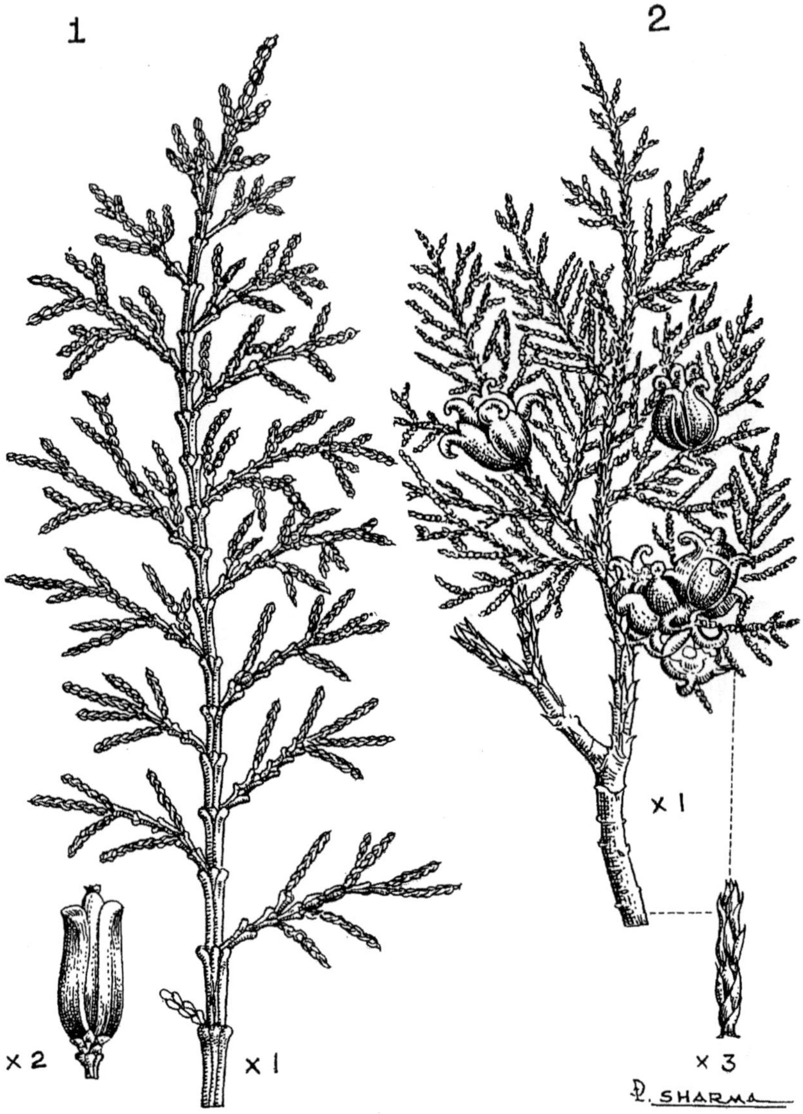

1. *Calocedrus macrolepis.* Branchlet and female cone. 2. *Thuja orientalis.* Shoot with female cones bearing downcurved hooks.

FAMILY 9 CUPRESSACEAE

PLATE 27

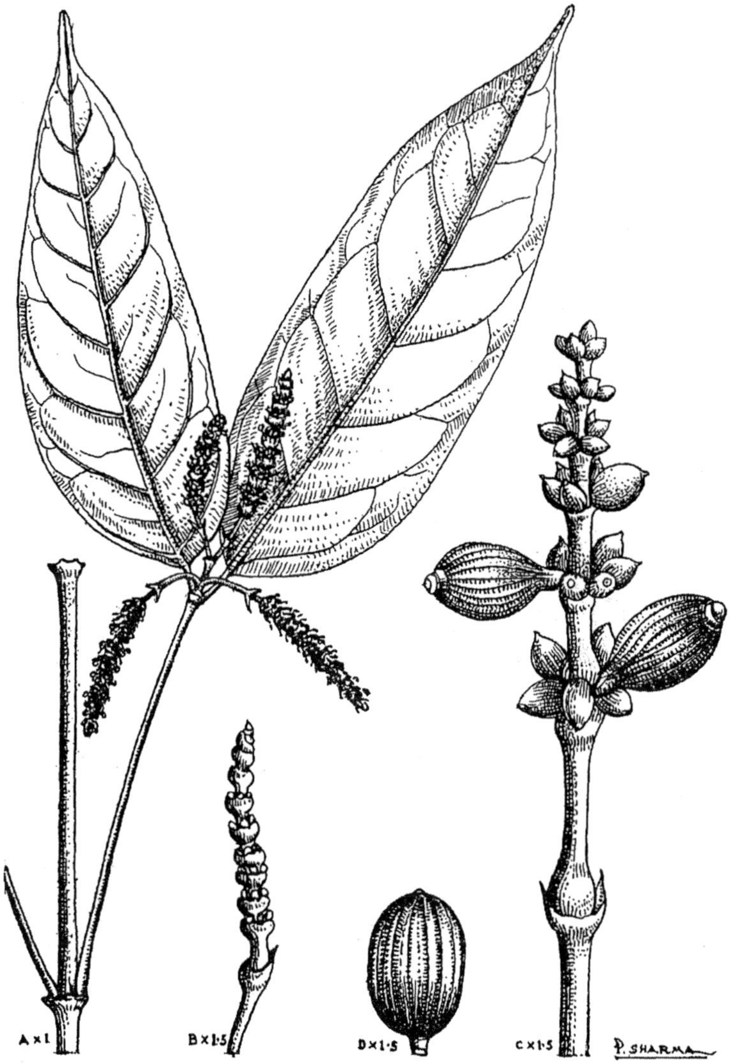

Gnetum gnemon L. from *Brandis 164* (DD). A. Branch with leaves and male inflorescence × 1. B. cones showing ovules × 1.5. C. Infructescences. × 1.5. D. fruit × 1.5.

FAMILY 10 GNETACEAE

PLATE 28

Gnetum gnemon L. var *griffithii* (Parl.) Mgf from *R.N. Parker* 2407 (DD) and *C. E. Parkinson* 1630 (DD) both from Burma. A. Branch with leaves and female cones × 1. B. Inflorescence showing female coness × 5. C. Disc showing female cones × 10. D. Branch with leaves and Male cones × 1.

FAMILY 10 GNETACEAE

PLATE 29

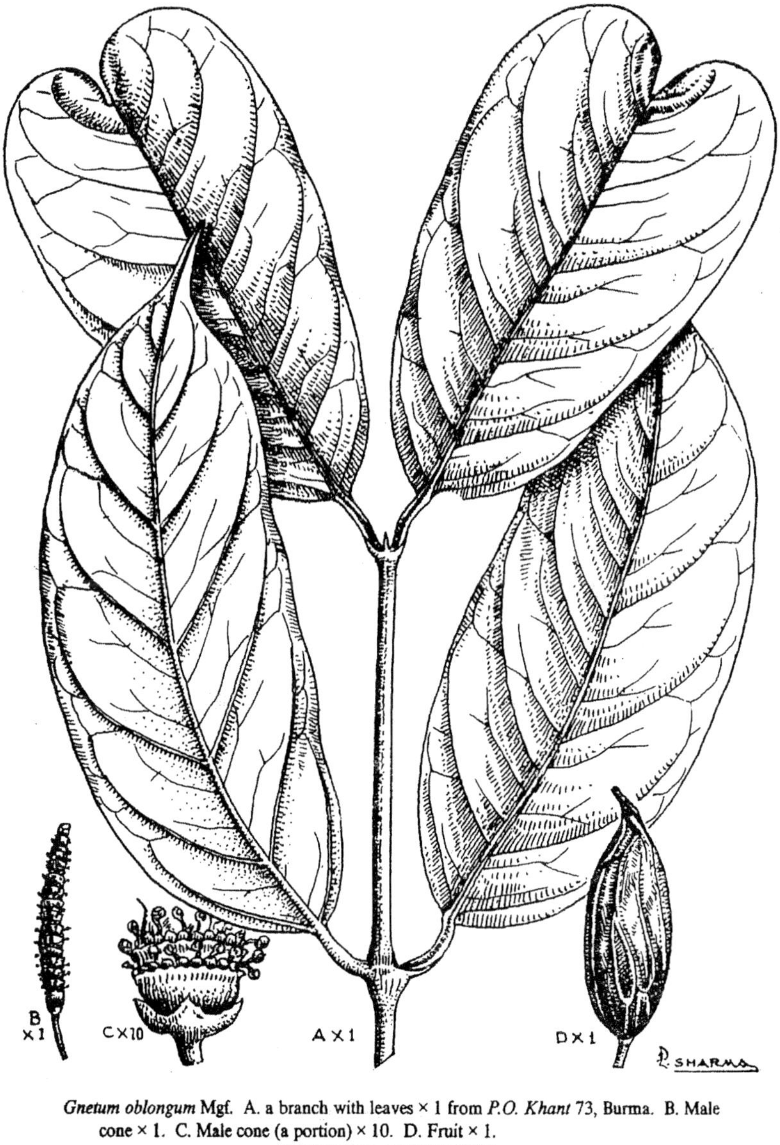

Gnetum oblongum Mgf. A. a branch with leaves × 1 from *P.O. Khant* 73, Burma. B. Male cone × 1. C. Male cone (a portion) × 10. D. Fruit × 1.

FAMILY 10 GNETACEAE

PLATE 30

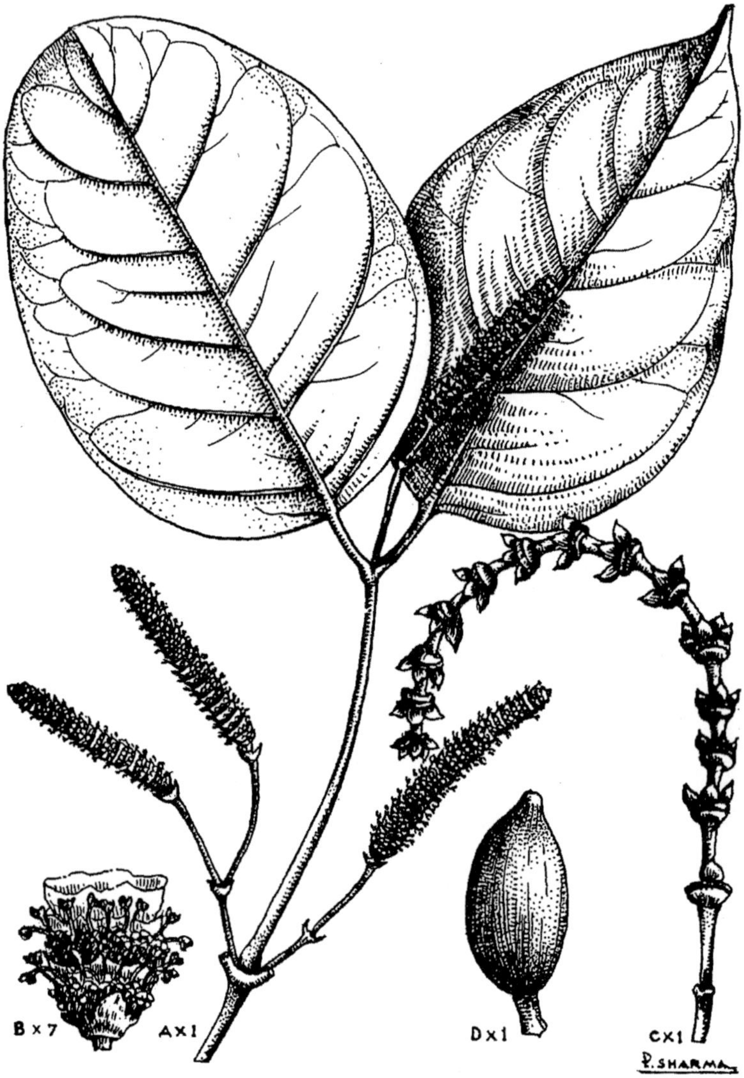

Gnetum ula Brongn A. a branch showing male inflorescence and leaves × 1. B. a disc showing male flowers × 7. C. a branch showing female cones × 1. D. fruit × 1.

FAMILY 10 GNETACEAE

PLATE 31

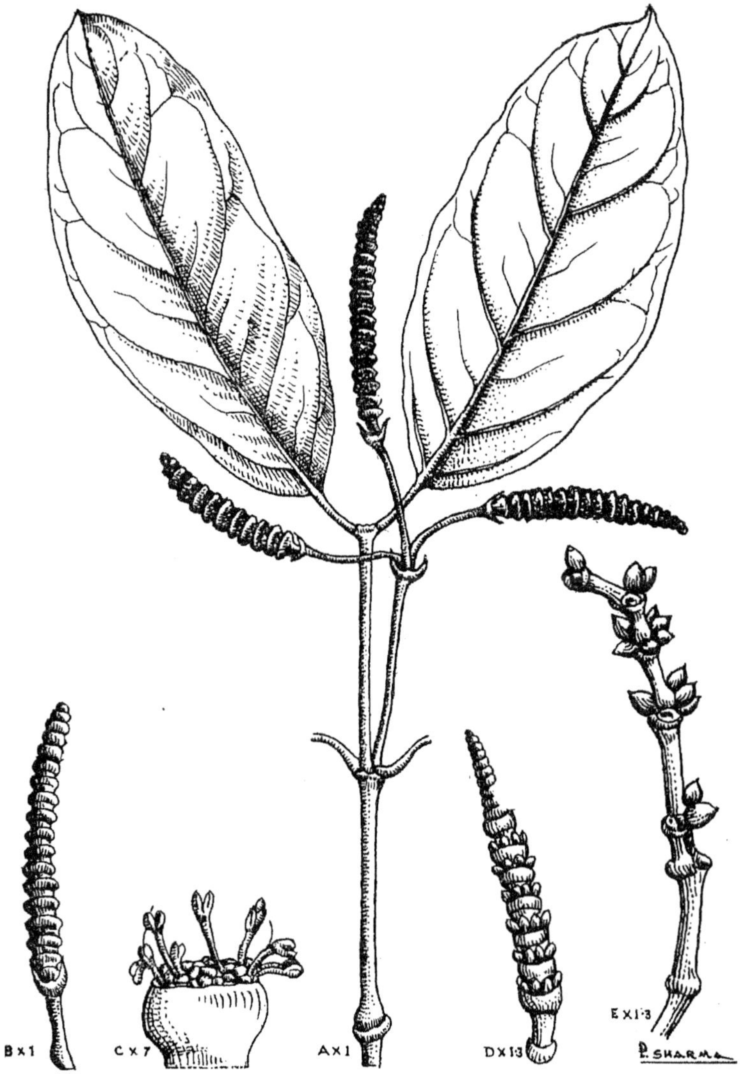

Gnetum ula Brongn. A. branch with male cones and foliage from *K.N. Subramanian* 2731 (DD) Kerala × 1. B. Male cone × 1. C. Male flowers × 7. E. enlarging and arrested ovules × 1.3.

FAMILY 10 GNETACEAE

PLATE 32

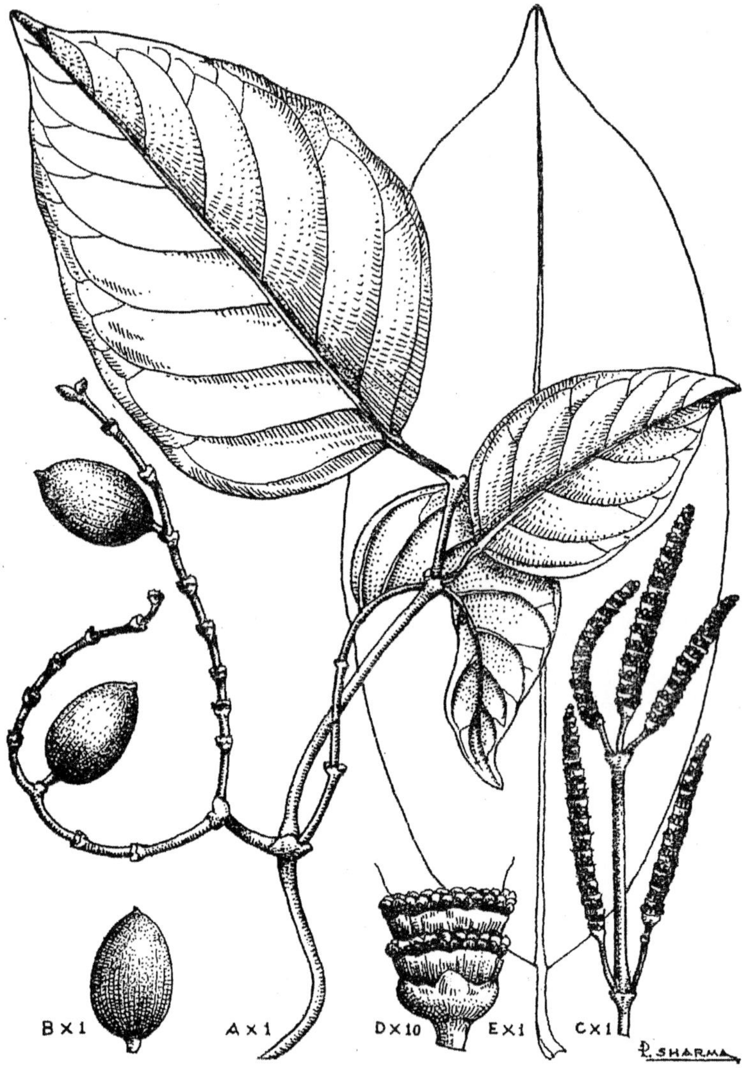

Gnetum montanum Mfg. A. shoot with leaves and female cones × 1 from *J.S. Gamble* 9814 (DD). Darjeeling. B. Female Cone × 1. C. Five male cone × 1. From *Rogers* 275 (DD). Burma. D. Male flowers × 10. E. outline of leaf × 1.

FAMILY 10 GNETACEAE

PLATE 33

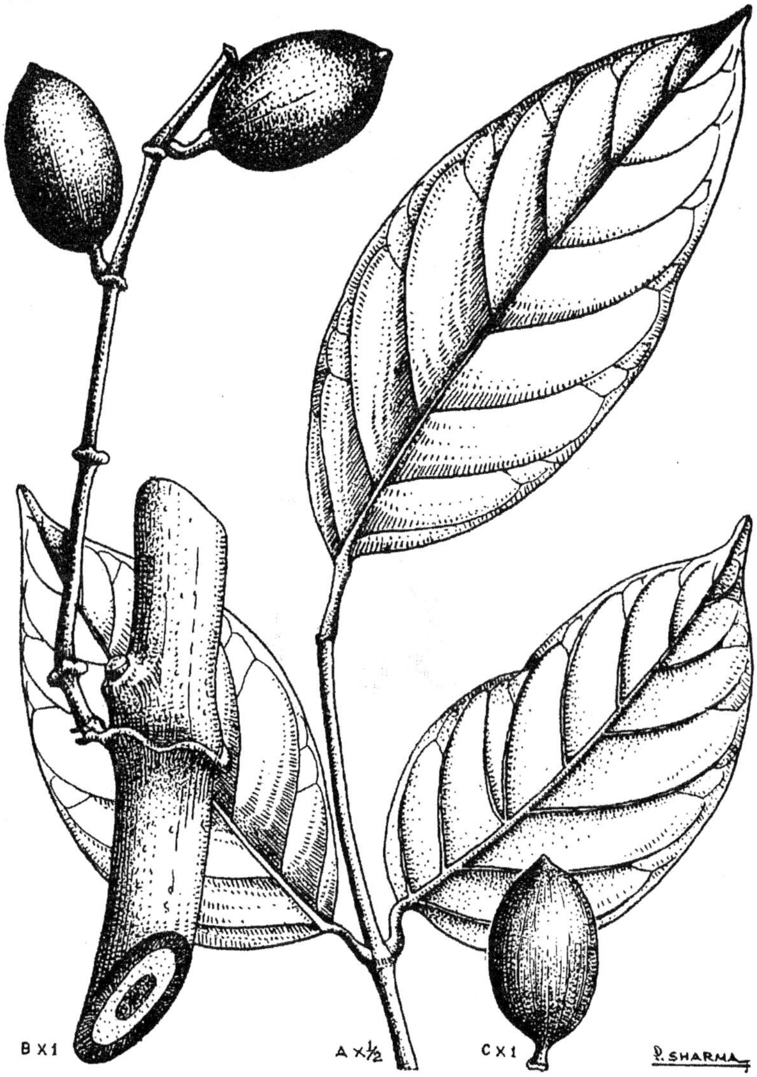

Gnetum montanum Mgf. f. *megalocarpum* Mgf. from *Lister s.n.* (CAL). Dafla Hills, Kameng Dist. Arunachal Pradesh. A. Branch with leaves × ½. B. Branch with female cones × 1. C. Fruit × 1.

FAMILY 10 GNETACEAE

PLATE 34

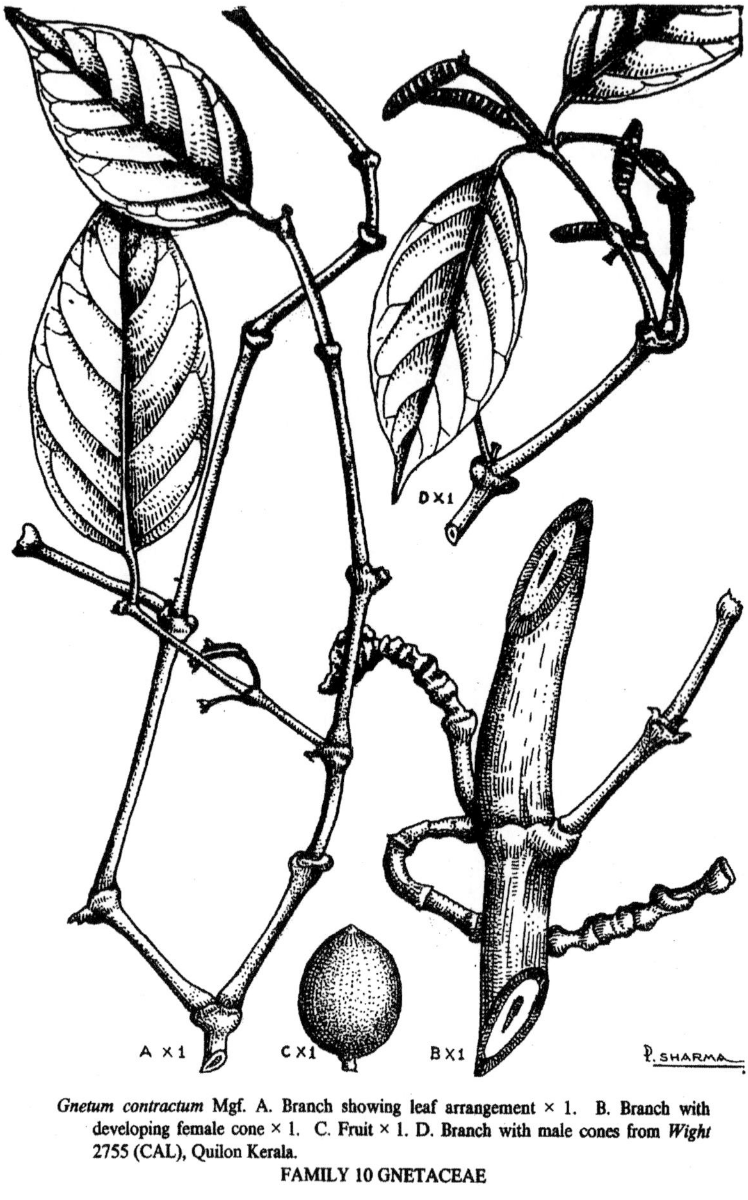

Gnetum contractum Mgf. A. Branch showing leaf arrangement × 1. B. Branch with developing female cone × 1. C. Fruit × 1. D. Branch with male cones from *Wight* 2755 (CAL), Quilon Kerala.

FAMILY 10 GNETACEAE

PLATE 35

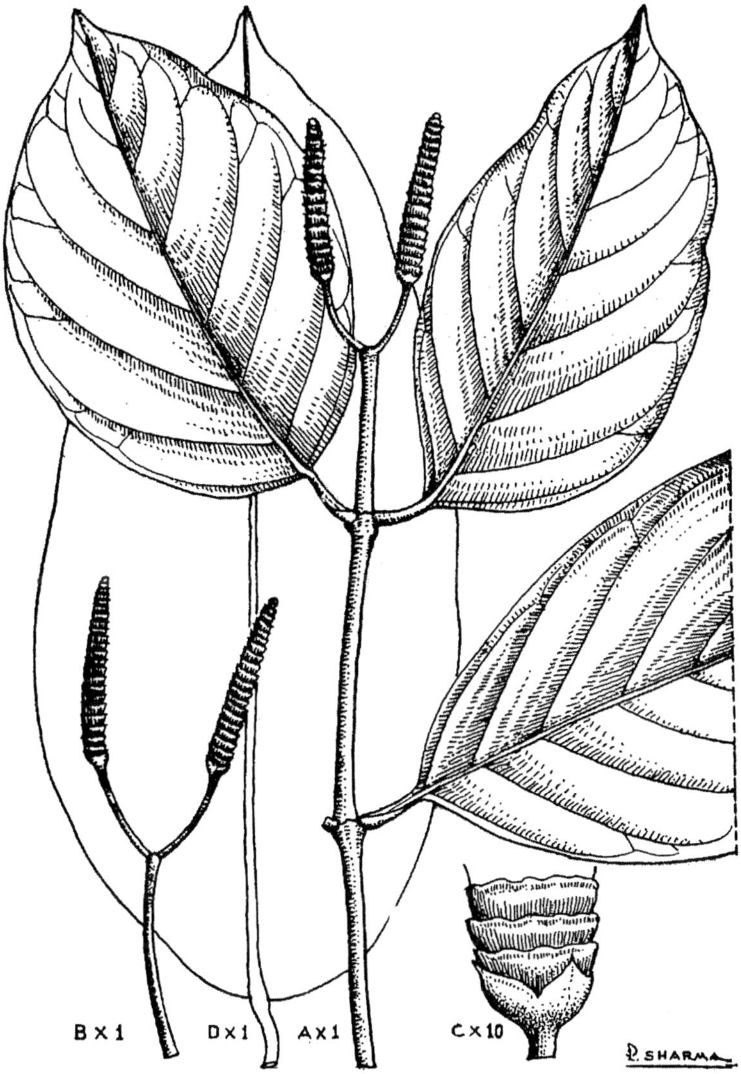

Gnetum latifolium Bl. var. *macropodum* (Kurz) Mgf from *C.E. Parkinson* 1067 (DD) S. Andamans. A. Branch with leaves and male spikes × 1. B. Twig with male spikes × 1. C. Male cone × 10. Leaf outline × 1.

FAMILY 10 GNETACEAE

PLATE 36

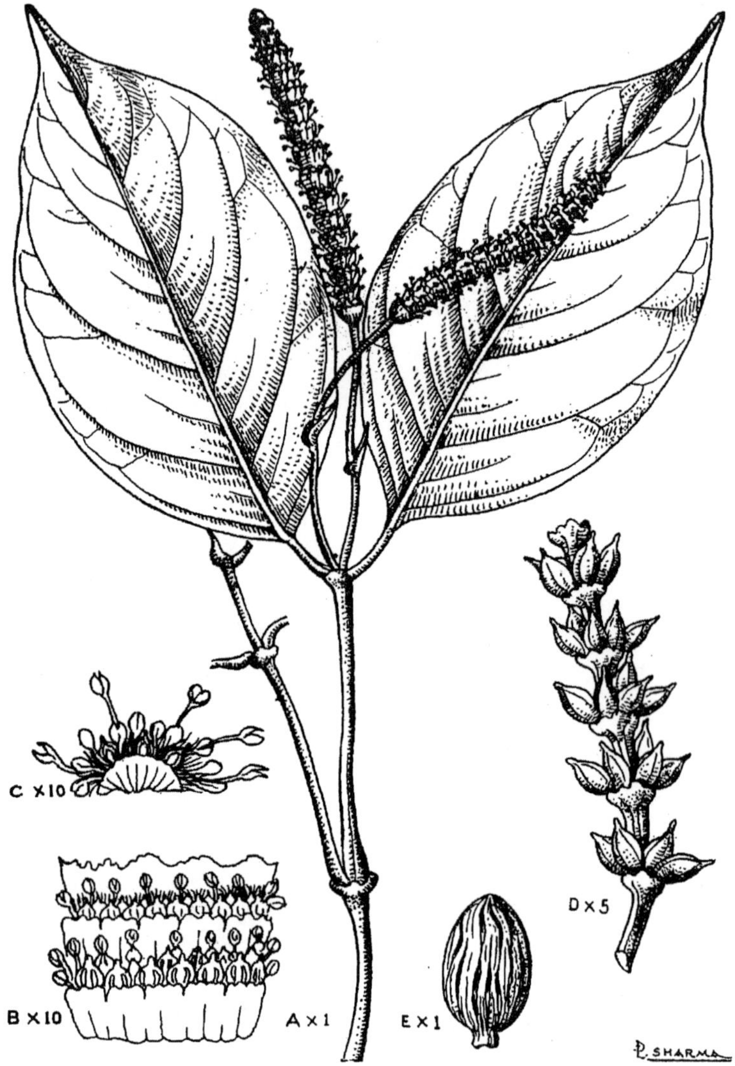

Gnetum latifolium Bl. var. *funiculare* (Bl.) Mgf. A. Branch with male flowers × 1. B. Disc Male flowers × 10. C. Male flowers × 10. D. Branch with female cones × 5. E. Fruit × 1.

FAMILY 10 GNETACEAE

PLATE 37

Gnetum macrostachyum Hk.f. from *R.N. Parker* 2284 (DD) Large climber Kalcinaung to Gebon (Burma). B. Fruit × 1. C. Branch with female cones × 1.

FAMILY 10 GNETACEAE

PLATE 38

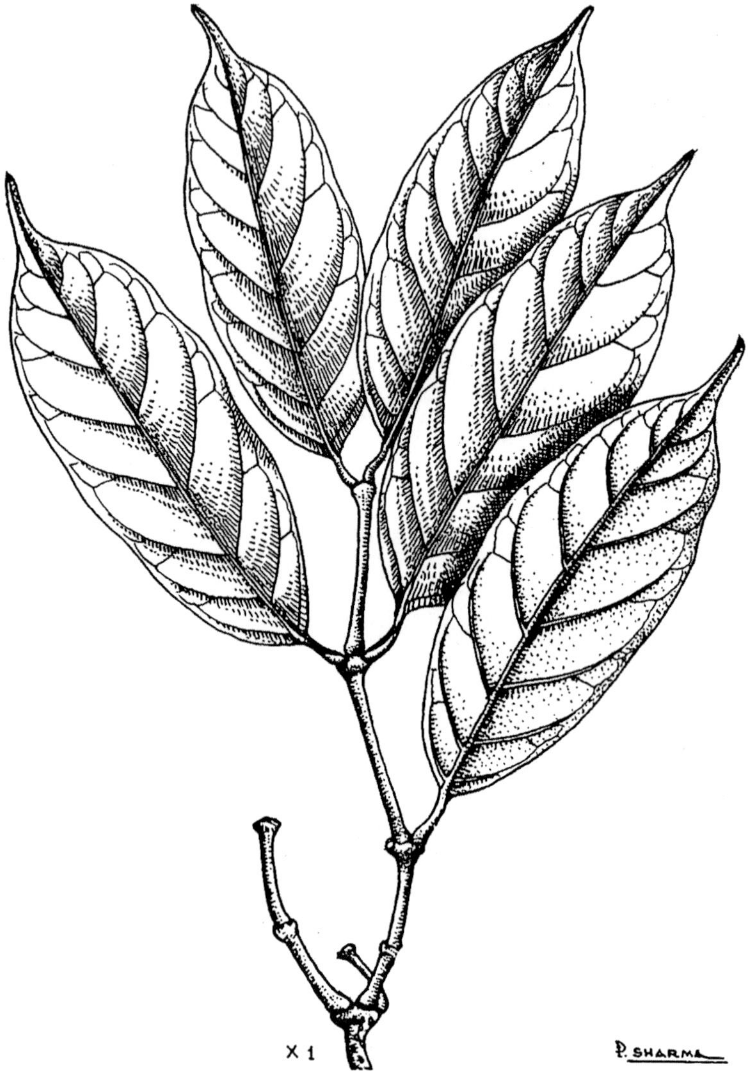

Gnetum microcarpum Bl. × 1. From *Griffith* 4972 (CAL) Mergui, Burma.

FAMILY 10 GNETACEAE

PLATE 39

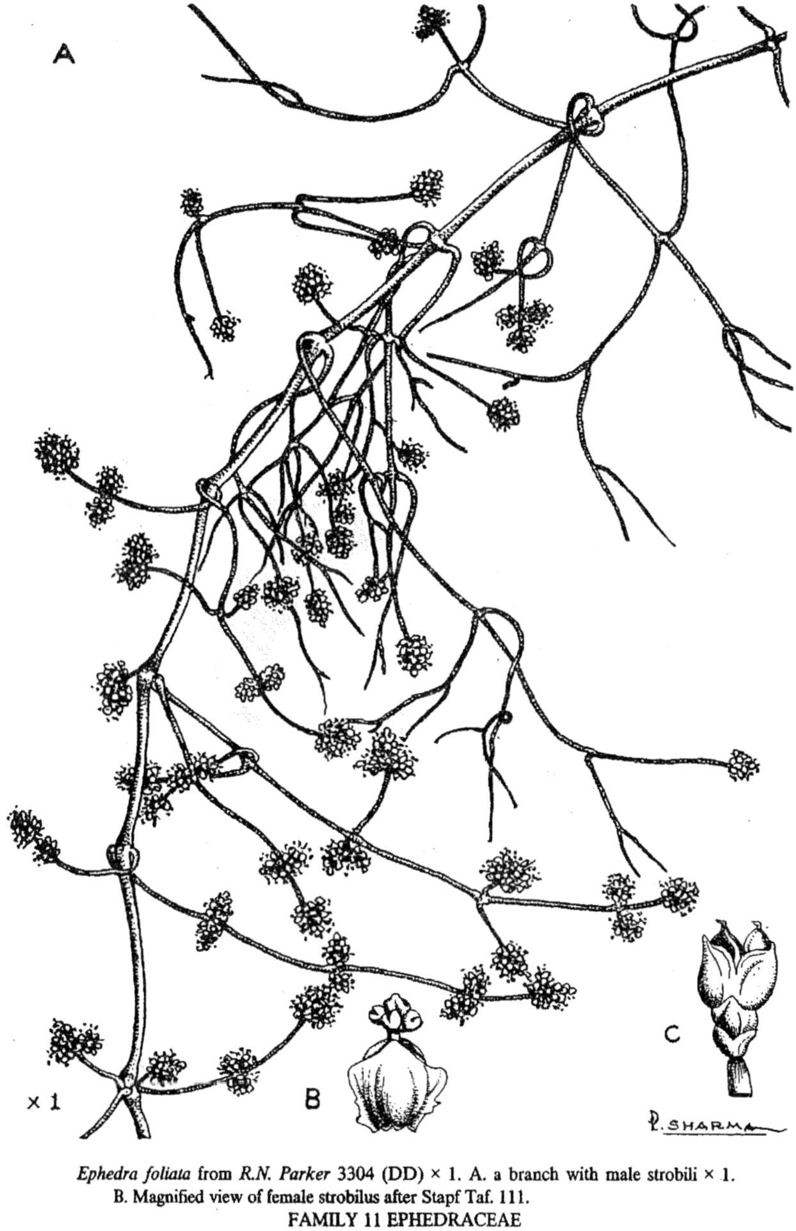

Ephedra foliata from *R.N. Parker* 3304 (DD) × 1. A. a branch with male strobili × 1.
B. Magnified view of female strobilus after Stapf Taf. 111.
FAMILY 11 EPHEDRACEAE

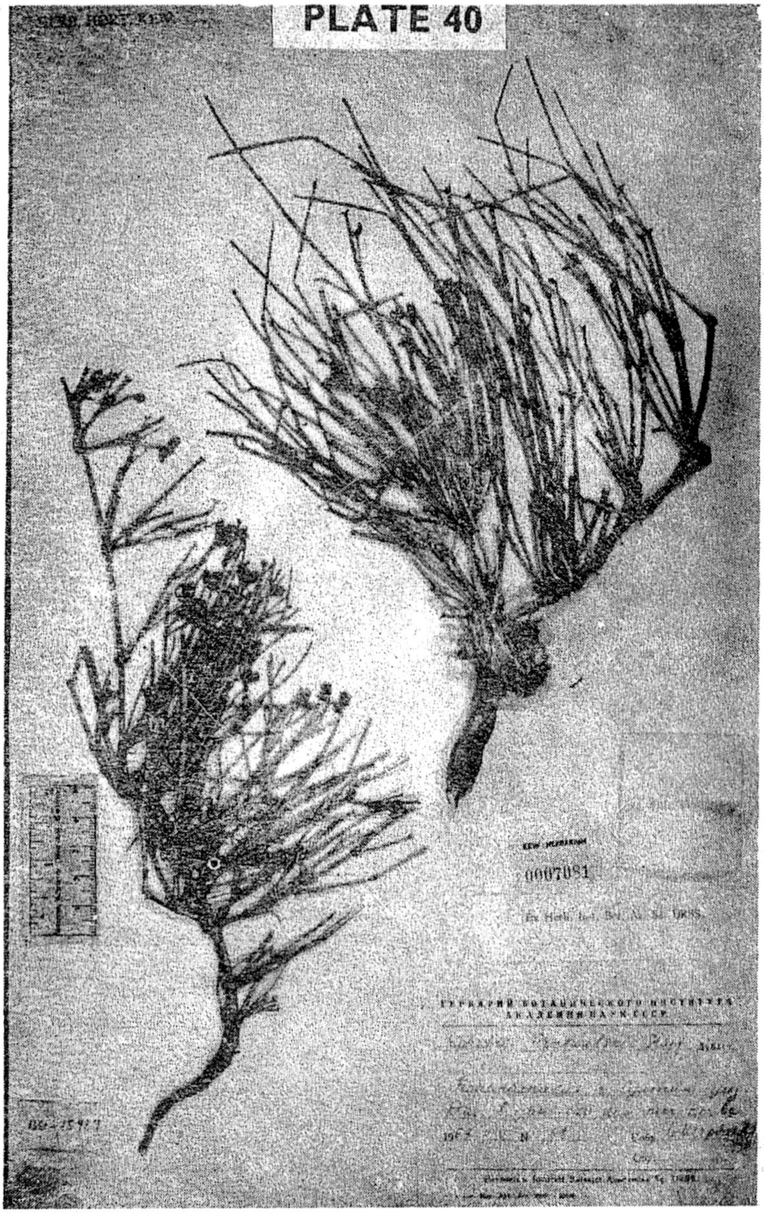

Ephedra przewalskii. Photograph of a specimen in Kew herbarium.

FAMILY 11 EPHEDRACEAE

PLATE 41

Ephedra pachyclada. A. Male twig × 1. B. a branch of flowers. C. a single flower enlarged. D. Female twig × 1. E. a single flower enlarged After Trans. Linn. Soc. Ser. 2. Bot. vol. III

FAMILY 11 EPHEDRACEAE

PLATE 42

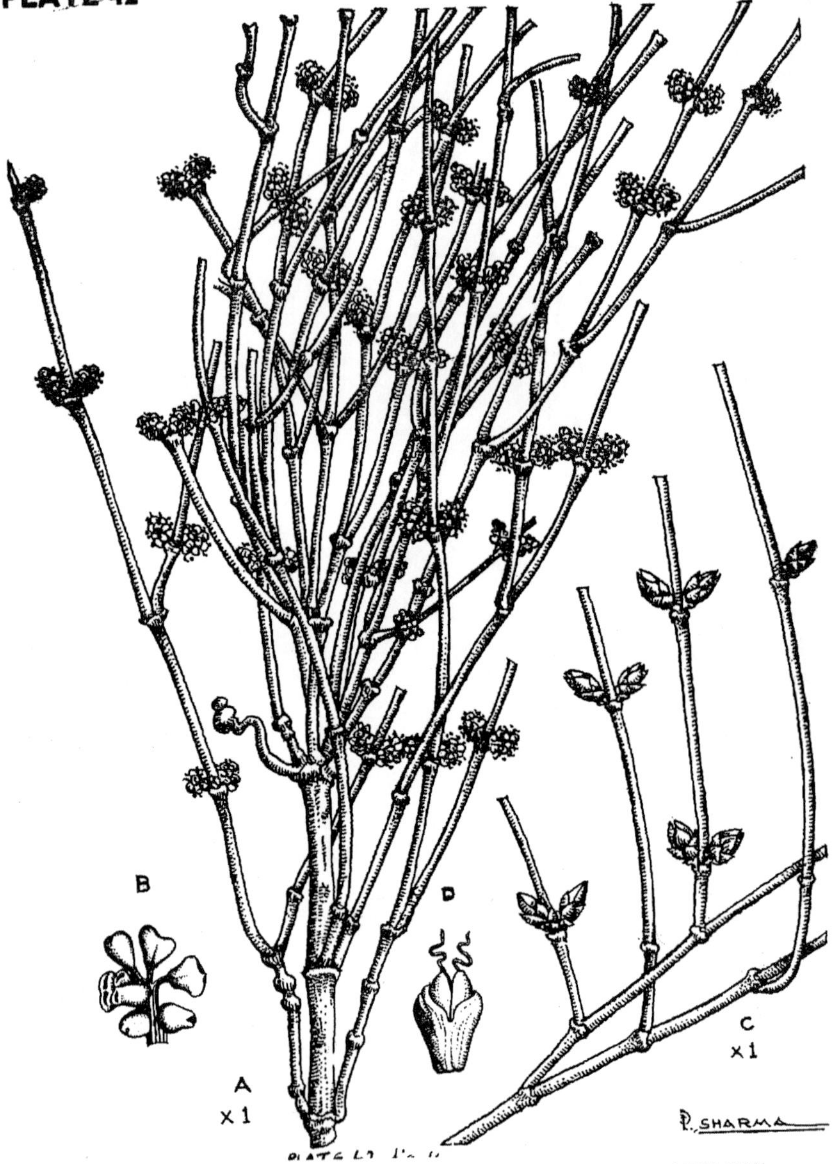

A. Ephedra intermedia var tibetica × 1. From *Lace* 1012 DD Khorangi Pangi 7000-9000′. B. Male strobili magnified. C. Branch with female strobili × 1. D. Female strobilus (Magnified). B & D after Stapf. Taf II (× 5).

FAMILY 11 EPHEDRACEAE

PLATE 43

Ephedra saxatilis. A. Branch showing general characters × 1. B. Branch with female strobili × 1. After *B.B. Osmaston* 5425. from a photograph After photograph Mus. Bot. Stockholm.

FAMILY 11 EPHEDRACEAE

PLATE 44

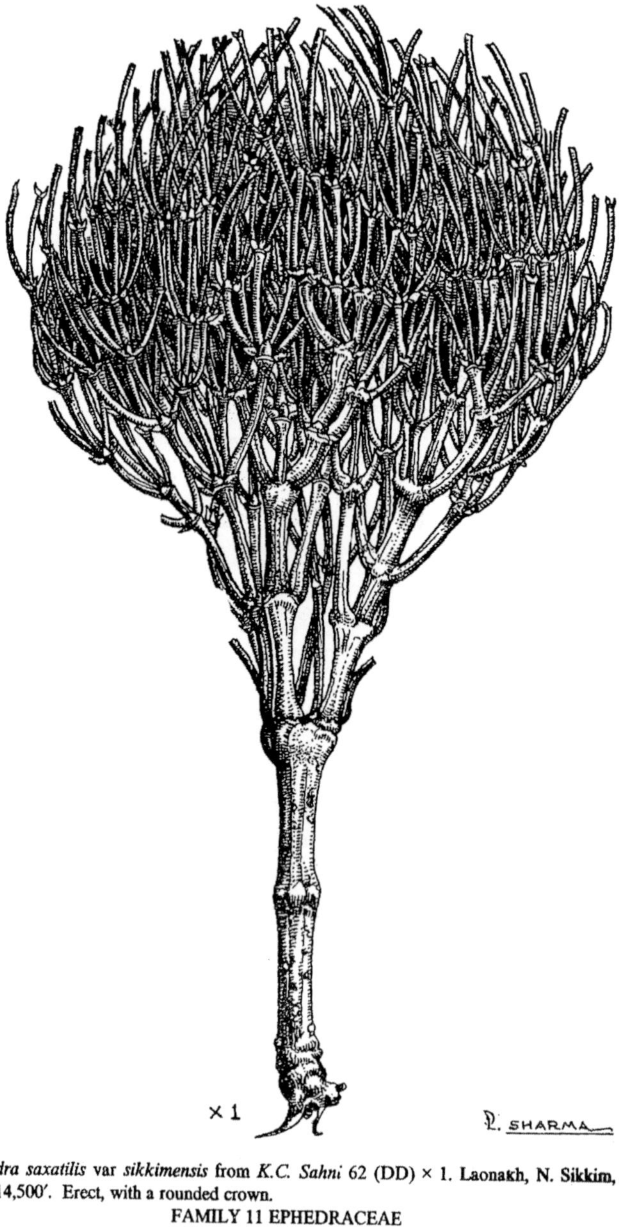

Ephedra saxatilis var *sikkimensis* from *K.C. Sahni* 62 (DD) × 1. Laonakh, N. Sikkim, 14,500′. Erect, with a rounded crown.

FAMILY 11 EPHEDRACEAE

PLATE 45

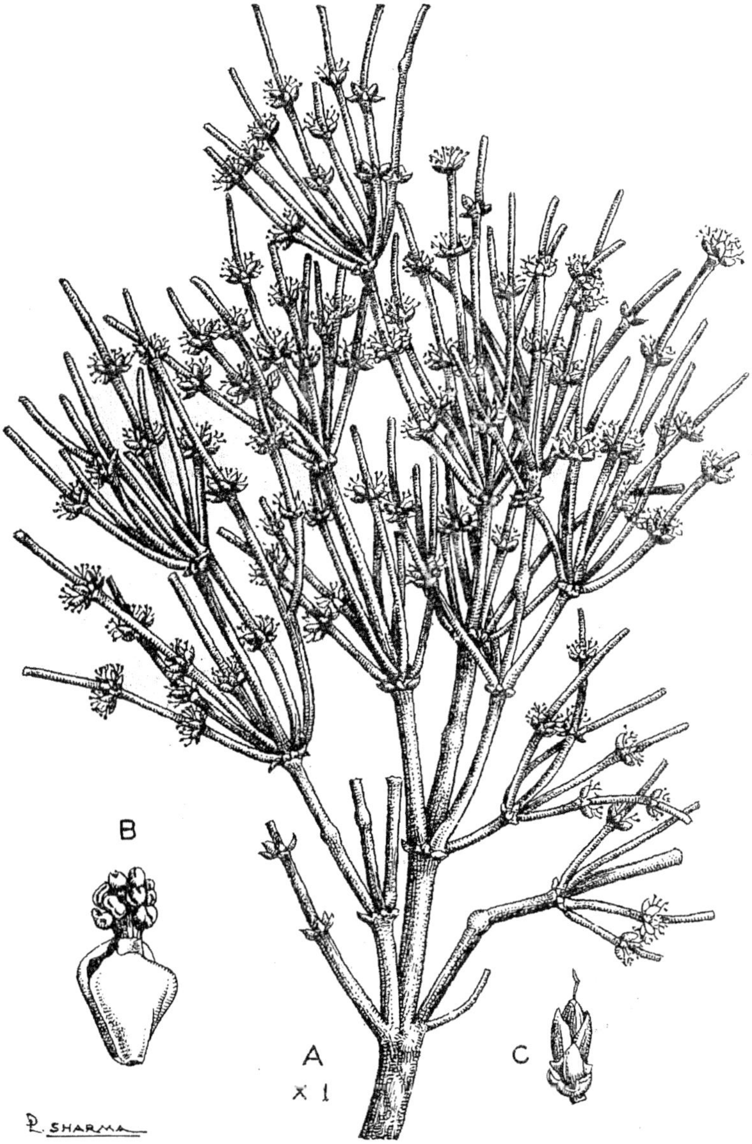

Ephedra gerardiana × 1 (From Herb spm). A. Branch with female strobilus × 1. B. Male strobilus, Magnified. C. Female strobilus. Magnified from Stapf Taf. III.
FAMILY 11 EPHEDRACEAE

PLATE 46

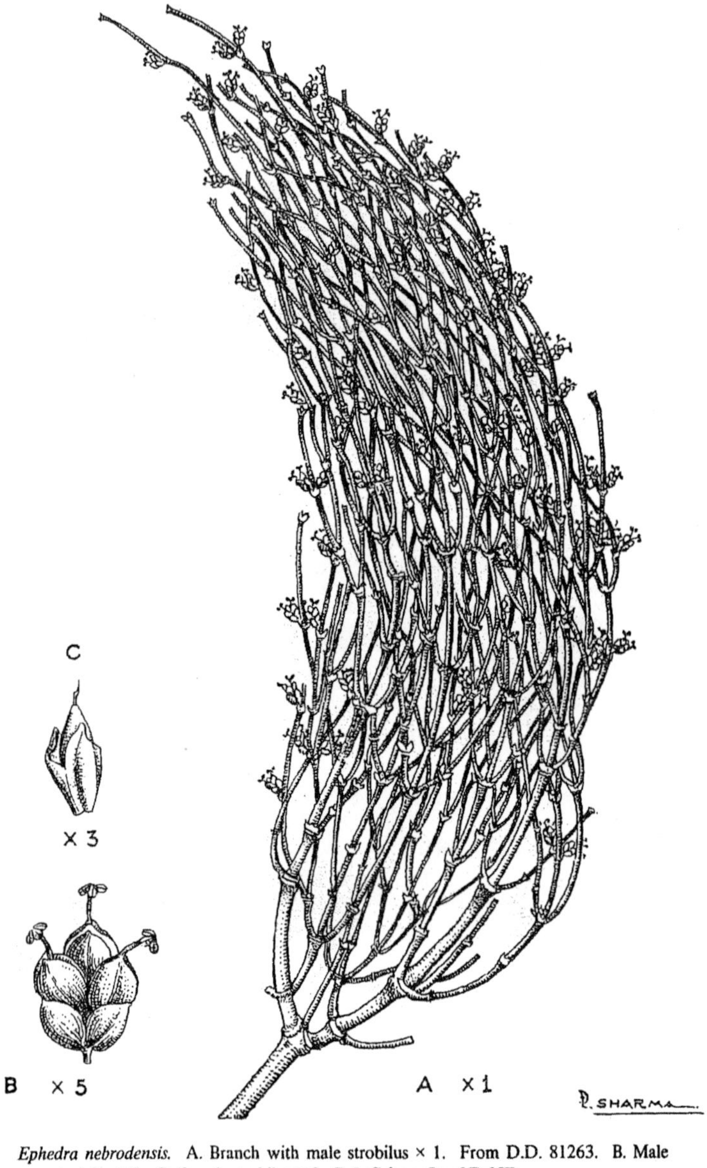

Ephedra nebrodensis. A. Branch with male strobilus × 1. From D.D. 81263. B. Male strobilus × 5. C. Female strobilus × 2. B & C from Stapf Taf III.

FAMILY 11 EPHEDRACEAE

PLATE 47

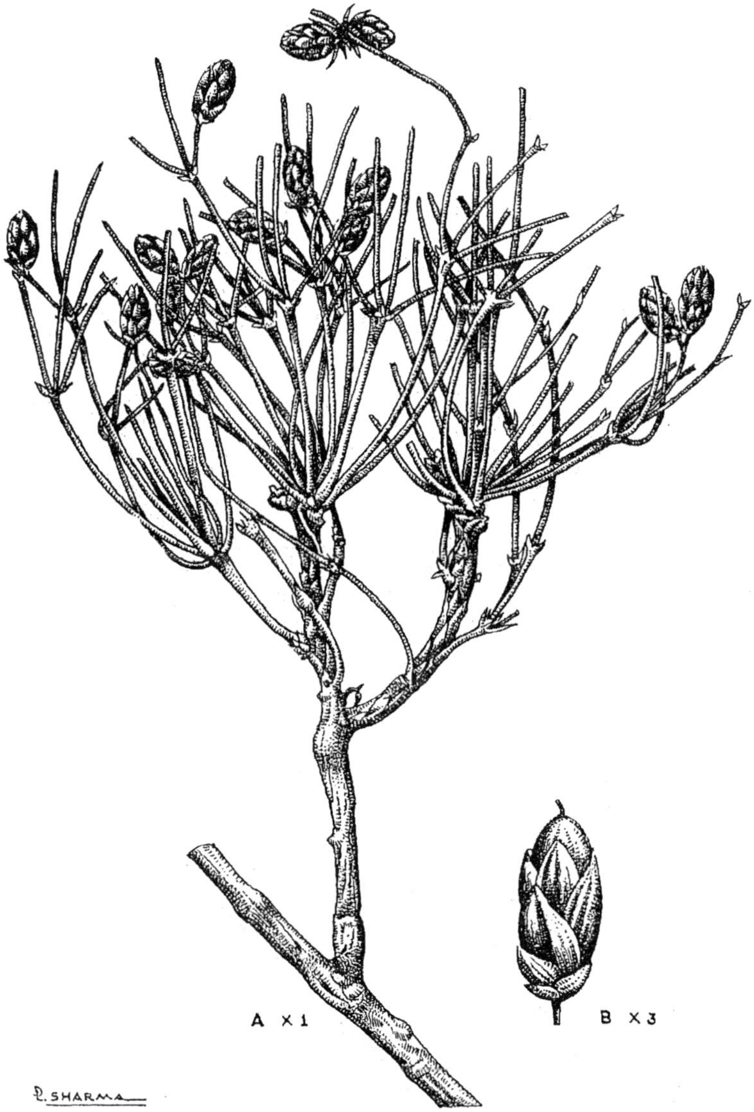

Ephedra regeliana. A. a branch with female strobili × 1. B. Magnified view of female strobili × 3. (After photograph Asia Media No. 4393 by V.J. Lipsky Naturhistoriska Riksmuseum Stockholm. (D.D. Herbarium).

FAMILY 11 EPHEDRACEAE

PLATE 48

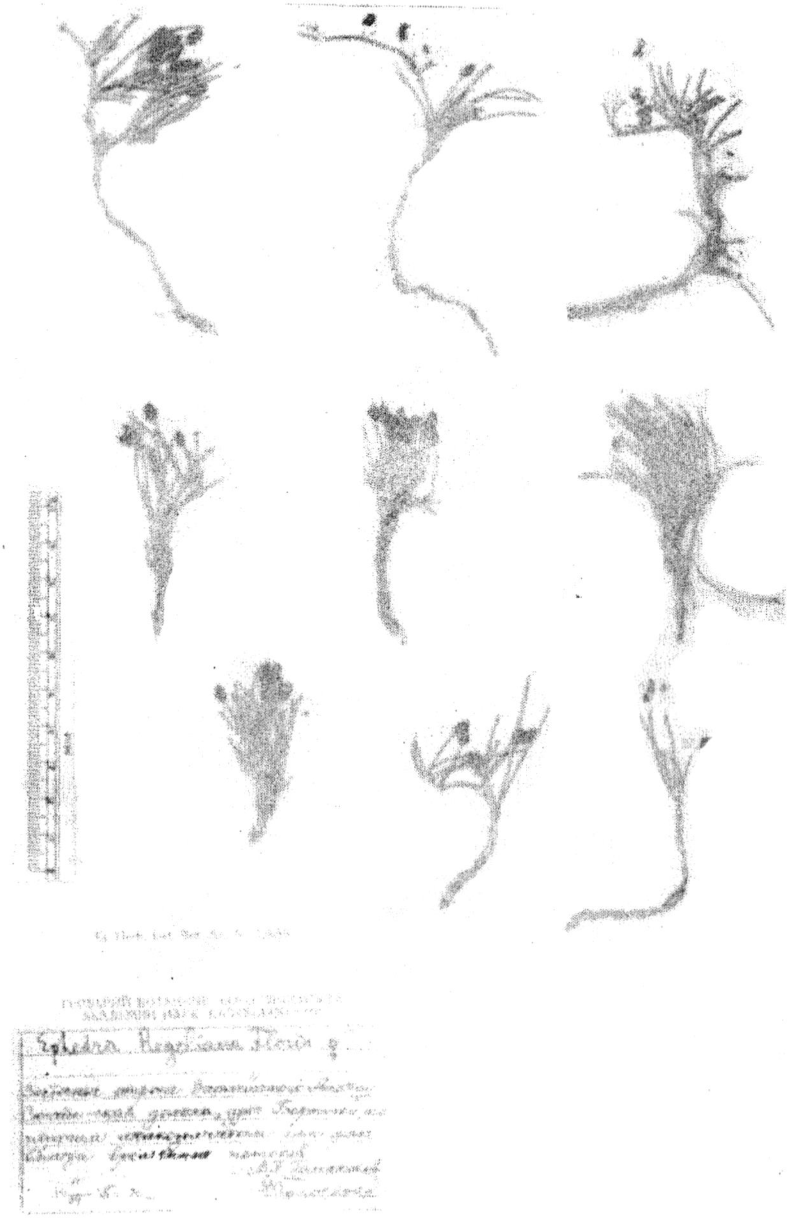

Ephedra regeliana Florin ♀ Photograph of a specimen in Kew Herbarium.

FAMILY 11 EPHEDRACEAE

1. Blue Pine, *Pinus wallichiana* Tehri Garhwal. © Vincent James.

2. Khasi Pine, *Pinus kesiya* Shillong. © FRI.

3. Fir forest (*Abies pindrow*) in Sonamarg Kashmir at 2000 m. Note columnar crowns of firs. Coniferous forests, given time regenerates itself crowding out broad-leaved intruders. This is well illustrated in the NW Himalaya as in this picture showing gregarious habit. © KC Sahni.

4. E. Himalayan Fir, *Abies densa*. Note flat branches. Umbrella-shaped. Singlila Range Darjeeling T. 9705. © FRI.

5. Himalayan Larch, *Larix griffithiana,* Yumthang Sikkim at 3000 m. At leaf-fall it is attractive with foliage turning a rich gold brown. In spring the new flush of leaves are fresh larch green and the young female cones are coppery red and look ornamental. From a photograph by KC Sahni.

6. Drooping Juniper, *Juniperus recurva.* 30′. Sikkim. In W Himalaya it is a shrub, in the E Himalaya a tree. After JD Hooker in *Himalayan Journals.* © KC Sahni.

7. Black Juniper, *Juniperus indica* 60′. Sikkim. In W Himalaya a shrub in the E Himalaya a tree. After J.D. Hooker in *Himalayan Journals*

8. *Cupressus corneyana.* Largest Cypress in the Himalaya. Ht. 120′. g.b.h. 18′. Rupa Kameng District. © K.C. Sahni W 22889.

9. *Cupressus corneyana.* Ht 30 m g.b.h. 8 m. Rupa, Kameng District. © KC Sahni.